FERMENTED

CHARLOTTE PIKE

CHARLOTTE PIKE is a graduate of the Ballymaloe Certificate Course and is a UK-based food writer.

This is Charlotte's fourth cookbook—*The Hungry Student Cookbook, The Hungry Student Vegetarian Cookbook* and *The Hungry Student Easy Baking* published simultaneously in August 2013 and won best series at the International Gourmand Awards in January 2014. Charlotte has contributed to a number of cookbooks by writing recipes, ghost-writing, recipe development, and food styling. Charlotte's recipes are simple, delicious, and really work. Charlotte is on the Committee of the Guild of Food Writers.

Always keen to share her knowledge, Charlotte's writing aims to inform and clearly instruct, and she's a lead tutor and Executive Chef at a small number of top UK cookery schools.

Charlotte has been writing a weekly blog about food for *HELLO! Online* for nearly 5 years and writes her own popular food blog, *Charlotte's Kitchen Diary*. Occasionally, she writes for other publications including the *Guardian* and broadcasts about food on BBC Radio stations, including BBC Radio 4.

Charlotte studied French and Spanish at Exeter University and then went to work in London before turning her love of good food and cooking into her career.

As a deeply passionate advocate of good food and excellent ingredients, Charlotte spends much of her time travelling the world learning about and searching for the best food and drink out there.

FERMENTED

A BEGINNER'S GUIDE TO MAKING YOUR OWN SOURDOUGH, YOGURT, SAUERKRAUT, KEFIR, KIMCHI, AND MORE

CHARLOTTE PIKE

PHOTOGRAPHY BY TARA FISHER

KYLE BOOKS

For T

Published in 2015 by Kyle Books
general.enquiries@kylebooks.com
www.kylebooks.com

Distributed by National Book Network
4501 Forbes Blvd., Suite 200
Lanham, MD 20706
Phone: (800) 462-6420
Fax: (800) 338-4550

First published in Great Britain in 2015 by
Kyle Books, an imprint of Kyle Cathie Limited

10 9 8 7 6 5 4 3 2 1

ISBN: 978 1 909487 37 6

Library of Congress Control Number: 2015941907

Editor: Vicky Orchard
Design: Lucy Gowans
Photography: Tara Fisher
Food styling: Annie Rigg
Props Styling: Tabitha Hawkins
Production: Nic Jones and Gemma John

Color reproduction by ALTA London
Printed and bound in China by C&C Offset Printing Co., Ltd.

CONTENTS

INTRODUCTION

Fermented food and drink is enjoyed in many cultures all over the world. Plenty of the foods we enjoy every day—such as coffee, chocolate, cheese, salami, olives, beer ,and wine—are fermented products. Fermentation is a naturally occurring process, in which carbohydrates are converted to alcohols and carbon dioxide or organic acids using either yeasts, bacteria or both, in the absence of oxygen. It has been used to preserve food as well as enhance its nutritional content for thousands of years. Every culture has its own fermenting tradition, from Japan's miso, China's kombucha and Korea's kimchi, to Middle Eastern labneh and Indian lassi, and the fermented pickles of cabbage, cucumber and carrots that feature prominently in the diets of north and east European countries from Bulgaria to Scandinavia. We can benefit from that combined wisdom by including some of these nourishing foods in our daily diet. It's easy to do and is far healthier and cheaper than buying supplements.

We all want to eat healthily, taking care to maximize our intake of fruits and vegetables and mostly that means using fresh, natural ingredients and avoiding refined sugar and carbohydrates. But something magical happens to certain foods, like cabbages and soybeans, that are allowed to steep in their juices so that their natural sugars and carbohydrates become agents to boost beneficial bacteria (probiotics) that our digestive tracts need to work best. The gut is part of our immune system and positively thrives when it contains good bacteria. They soothe and heal many digestive problems such as IBS, and can boost our general well-being by improving our immunity, helping with weight loss and even brain health, by reducing depression and anxiety.

Perhaps you've picked up this book because you've heard of fermentation, but you don't really know what it is. Maybe you've heard about the health benefits and you want to introduce these highly nutritious foods and drinks into your diet. Or perhaps you'd like to try something completely different in the kitchen. You need hardly any equipment or space to get started—just a few jars or bottles, some string, and cheesecloth are the most-used items. They are easy and inexpensive to acquire, and won't take over your kitchen.

Although some recipes take time to ferment, your hands-on work is minimal. I have included lots of delicious recipes that make use of your fermented creations to help incorporate it into your diet every day. Once you get going, you'll find a natural rhythm of fermenting that works around your schedule. It won't take over your life; it will fit quietly in around it, producing delicious and nutritious rewards. If you want to read about the science surrounding fermentation, there are other excellent books that will give you all the detail you require. I am a cook, not a scientist, and my intention here is to simplify the science and processes, and translate that into foolproof recipes that anyone can successfully achieve.

This book is the result of rigorous experimentation and I have included only the easiest recipes so that you can get consistently good results right from the start. Not many of us have the space, time, and inclination to spend days and weeks following lengthy processes to make something that is, frankly, just as delicious—if not cheaper—than fermented foods purchased from an artisan producer. So, I want to show you what you can very successfully make at home, but I will also advise which fermented foods are best bought from the small number of great producers out there (see page 159 for suppliers).

My journey into fermentation started while I was studying at Ballymaloe Cookery School in Ireland. Darina Allen, her daughter-in-law Penny, and other teachers at the school are highly committed to revisiting and reviving forgotten culinary skills and passing on their wisdom. Fermentation is one aspect of producing and eating food that gives maximum health benefits that they teach at the school.

Fermentation had only been vaguely on my radar until then, and my experiences at Ballymaloe spurred me on to look at incorporating these foods into my diet in a realistic way. Keen to try a whole range of ferments, I quickly started producing jar upon jar of delicious ingredients for my pantry and fridge, and I wanted to find ways of incorporating what I'd made into my every day cooking. My research and experiments took me around the world, and resulted in many delicious discoveries that I now enjoy daily.

The benefits of including fermented foods in our diet are numerous. The process of fermentation makes the raw ingredients easier to digest and their nutritional content readily available and therefore more easily absorbed by our bodies. Fermented foods deliver those beneficial bacteria (probiotics), enzymes, nutrients and detoxifiers that our immune systems and digestive tracts need. Instead of buying supplements, by eating and drinking fermented foods, over time you will see a real improvement in your overall health and well-being—physically, mentally and emotionally.

I hope you will enjoy these recipes, and discovering the process of fermentation. It's much easier than you'd think. And when you get into the swing of things, it's something that stays with you. For life.

NOTES ON INGREDIENTS

As with all cooking, using the best quality raw ingredients will result in the best end product. This is particularly so with fermenting, because additives can affect the fermentation process.

A good general rule of thumb is to use organic ingredients when you are fermenting. Later, when incorporating non-fermented ingredients into recipes, organic is preferable, but not essential, especially if you have difficulty sourcing organic ingredients.

WATER

Pure water is essential for fermenting. Any chemicals such as chlorine in the water will not allow your food to ferment as it should. If you're on municipal water, never use water straight from the faucet, as the chlorine present in the water supply will inhibit the fermentation process. You can use pure bottled mineral water, or for a more economical and ecological option, pour tap water into a large, wide, shallow bowl and leave it to sit, uncovered, overnight. The chlorine will evaporate and the water will be good to use by the morning.

Filtered tap water is fine to use, but again I recommend you leave it overnight as mentioned above for unfiltered tap water.

The quality of the water you use is crucial for making fermented drinks such as Water Kefir (page 126) and Kombucha (page 133). Water is lacto-fermented by adding water kefir grains, a live, active culture. They look like small, gelatinous crystals, and are little structures made up of bacteria and yeast that feed on sugar and convert it into beneficial bacteria, enzymes, and nutrients. There is no need to rinse the kefir grains between uses.

Lacto-fermentation is the process of fermentation brought about by the *Lactobacillus bacterium* (named so because they were first discovered in milk, even though this form of fermentation is not restricted to milk products). *Lactobacillus* bacteria convert sugar into lactic acid, which inhibits harmful bacteria from growing, and increases the vitamin and enzyme levels in food and drink, as well as boosting digestibility.

Kombucha is made using a scoby, which is an acronym for Symbiotic Colony Of Bacteria and Yeast, another culture; a living home for bacteria and yeast, like water kefir grains, but which looks totally different. Often a light brown color with tinges of orange, a scoby looks a bit like a strange gelatinous pancake, often with brown strands of yeast growing from it. It will grow over time, and layers can be peeled off to make new batches.

Scobies tend to thrive when given a rich dose of minerals. If you're fortunate enough to have a supply of freshly drawn water at home, you may notice your fermented drinks flourish particularly well. That's not to say your ferments won't work using municipal water, provided you have allowed the chlorine to evaporate off, but you may like to give your drinks a boost by using pure mineral water (say one in every four ferments).

MILK

Raw, organic, unpasteurized milk is the number one choice for making yogurt (see page 44) and milk kefir (see page 130)—and you're very lucky if you can source it. Most of us can't find raw milk so readily, in which case I recommend you select organic, un-homogenized milk, which is available in the more high-end supermarkets and health food stores. Whole milk works best and gives the most flavor. If you are making labneh or using store-bought yogurt, I strongly recommend using organic, as with milk.

Milk is fermented into milk kefir using milk kefir grains, which are different to water kefir grains. They are smaller and resemble a sort of jellied cauliflower. As with water kefir grains, please make sure you don't rinse them between uses.

VEGETABLES AND FRUIT

In order for your ferments to work, your ingredients need to be pure, natural, and additive free.

Organic vegetables and fruit are preferable again, when they are the raw ingredients that are being fermented. Non-organic is fine for recipes that aren't actually fermented, such as stir-fries and pancakes.

Before using vegetables and fruit, it is sensible to give them a quick rinse first. Do take the time to check the condition of your produce, too; cut out or discard any bruised pieces, as they may introduce unwanted mold to your ferments.

Always make sure your fruit is ripe—avoid under or over-ripe food for fermenting. Use it in other ways, such as jam making (under-ripe) or cake baking (over-ripe). If you're using dried fruit, it's always a good idea to select organic, unsulfured fruit.

SUGAR AND HONEY

Where sugar is required, please seek out organic, unrefined superfine sugar, which is usually cane sugar. Honey must be raw because it is in this natural state that honey offers most health benefits. Raw honey is rich in enzymes and antioxidants because it is in its natural state when it has not been heat-treated.

SALT

Unrefined, pure sea salt is an important ingredient to have on hand for fermenting. It is important because many refined salts contain added iodine, which kills the crucial microorganisms that cause the ferment. I use Maldon sea salt and it works well, producing a delicious result both in flavor and texture.

NOTES ON EQUIPMENT AND SAFETY

STERILIZING JARS

In order to ferment successfully, your equipment must be scrupulously clean. In terms of cleaning and preparing my equipment, I follow the same procedure as I do with preserving; I like to run my containers through the dishwasher on a hot cycle and use them shortly afterward. Hand washing is fine as long as you use dish soap and plenty of hot water and rinse well. You may wish to rinse your equipment and vessels again after removing them from the dishwasher, just in case the detergent hasn't been removed effectively by the rinse cycle.

CONTAINERS

When producing any ferment, always use glass or ceramic containers—never metal or plastic, which will react with the ferment. I only use Le Parfait rubber seal clip-top jars and bottles. I keep a variety of different sizes, but I find the 1-, 2- and 3-liter sizes are the most versatile.

You can buy pottery crocks for fermenting kombucha, however I prefer to use a 3-liter Le Parfait jar and leave the lid off, covering the surface with a piece of cheesecloth tied with string. This is a cheaper and slightly more space-saving option.

It's always worth labeling your containers—not only with the contents, but also the date, so you can keep an eye on how your ferments are progressing.

CHEESECLOTH

Cheesecloth is essential for making labneh (strained yogurt, see page 64). It also makes a good covering for ferments, as it protects them from flies and dirt if you are leaving the top off, and allowing air to circulate. I usually tie the cheesecloth in place using string. Cheesecloth is widely available from cooking stores and online.

You could use a pure cotton dish towel or cloth as an alternative. Just make sure it's really clean before using: wash on a high temperature and dry thoroughly before use.

BANNETONS

A cane banneton basket is a nice piece of equipment to use when making sourdough bread (see pages 104–109). It provides crucial shape and structure to a soft, wet sourdough, and produces a very pretty pattern on the dough. If you use one, make sure it is well floured. A mixture of 50 percent white bread flour and 50 percent rice flour is best to avoid the dough from sticking, but otherwise, all purpose or plain white bread flour will do. If you don't have a banneton, use a medium-sized round mixing bowl. Liberally coat a clean dish towel with flour and place this inside the bowl before laying your dough on top.

WHERE TO LEAVE YOUR FERMENTS

Fermentation mostly takes place outside the fridge. Every home is different, so I suggest finding both a warm and a cool, dark place for ferments. Many are quite happy being left out on a countertop, some need to be kept in the fridge, such as anything dairy-based, and others need a cool, dark place. You don't need a lot of space, and the kitchen countertop away from direct heat is an ideal place for many ferments.

SAFE FERMENTING

Fermentation is generally very safe, and with clean equipment and good quality ingredients, the risks are very low. Everyone's ferments work differently, though, and if you find mold forming in or on yours, then something has gone wrong. Unfortunately, you will have to throw your ferment away and start again. Mold is generally caused by poor storage conditions, poor hygiene, or the food not being properly submerged before fermenting. Any surfaces that are floating or not covered in liquid will turn moldy, so need to be submerged fully. Consult the USDA guidelines for food safety guidance and best practice when canning and fermenting foods (nchfp.uga.edu).

Some people worry that fermenting may lead to their containers exploding. This can happen with some ferments, such as beer making, but all the recipes in this book are very safe and highly unlikely to explode. You may see bubbles forming in your jars, which is totally normal. That said, if you feel as though the pressure is building in your containers, simply ease off the lid, gently and carefully, to let the air out, then close the lid again. This is called "burping," and can be done as regularly as you feel is necessary.

It's really important to ensure your produce is free from bruises or mold before starting to ferment. Bruises will most likely lead to mold forming on your ferments, and starting with mold is a no-no.

STORING YOUR FERMENTS

If you're going away, or just having a break from fermenting, your key ingredients—the sourdough starter, scoby, and kefir grains—will keep well. Simply place at least 3oz (75g) of sourdough starter in the fridge in a sealed plastic or glass container. Over time, a layer of dark liquid will form on top. This is completely normal. It's called hooch, and is a protective layer of alcohol and vinegar, which will keep the sourdough dormant. When you are ready to use the sourdough starter again, you can either stir in the hooch, or pour it away and refresh the starter to make the levain (see page 108).

The scoby can be kept in a sealed jar of kombucha for up to a month. When you are ready to use it, pour the kombucha away—don't drink it—retaining 1 cup to start the next batch.

Water kefir grains keep well in a sealed container in the fridge, too. Simply pour enough water over the grains to cover and add a couple of tablespoons of sugar. Secure the lid, shake, and leave in the fridge. To reactivate, pour away the liquid and use the grains in a fresh batch of water kefir.

Likewise, milk kefir grains can be stored in a sealed container in the fridge. They will keep, submerged in milk, for up to 14 days. If you need to keep them for longer, I recommend giving them to a friend to look after for you who will be able to refresh the milk they're submerged in weekly. To reuse your grains, pour away the liquid and start with a fresh batch of milk kefir.

FRUIT & VEGETABLES

SAUERKRAUT

MAKES 1 × 1-QUART JAR

1 lb organic white cabbage, very thinly sliced or grated
4 teaspoons sea salt
Water Kefir (page 126—optional)

you will need a 1-quart glass Le-Parfait-style jar with a rubber seal, sterilized according to the instructions on page 11

TIPS:

Do try to use organic cabbage if you can find it. Non-organic may contain chemicals that could interfere with the fermentation process.

It is important that the cabbage is really tightly packed in the jar. You will need to use a little force to press everything in.

Do ensure the cabbage is completely submerged in the jar. It needs to be completely covered in liquid to ferment.

This is probably one of the most well-known fermented foods in the world. My version is easy to make and can be ready to eat in less than a week. The vegetables retain their crunch and don't have the vinegary liquid that is ever-present in so many store-bought varieties. Sauerkraut makes a delicious accompaniment to a wide range of dishes, and a spoonful is an excellent addition to many salads and sandwiches.

Place the cabbage and salt in a large glass or ceramic mixing bowl and toss together using your hands so that the ingredients are really well combined. Gently squeeze the cabbage, to encourage it to start to release some of its water.

Pack into a 1-quart jar, pressing down well so that the cabbage releases more water. The cabbage should now be sitting under a small layer of water. If your cabbage doesn't release much water, you can top it up with water kefir. It needs to stay completely submerged to ferment and not go bad, so you may need to put a small dish or weight (glass or ceramic only, not metal) to sit in the top of the jar, leaving the lid off for a couple of days, to get it started. If fermentation doesn't happen instantly, the cabbage will stay submerged after a couple of days weighted down. Fasten the lid as soon as the cabbage stays submerged unaided—this could be right away, or after 2-5 days.

Set aside in a cool, dark place for 1 week to ferment, after which it will keep, unopened, for up to 3 months. It will turn a light brown in color as it matures. Once opened, store in a cool, dark place and eat within a month.

SAUERKRAUT WITH FENNEL AND CARAWAY

Add 1 grated bulb of fennel and 2 teaspoons of caraway seeds to 5 cups grated white cabbage and 4 teaspoons of salt, following the above method. This is also delicious with a grated apple stirred in just before serving.

KRAUT SLAW

This delicious salad is a great way of using homemade sauerkraut and introducing it into every day cooking. By stirring the sauerkraut into raw vegetables to make a slaw, you retain the crunchy texture and maximum nutrients from the ferment.

Serving sauerkraut in this way is a great way to introduce people to its flavors. It can be quite strong to eat on its own and you will probably want to eat a only little at a time, but this slaw is a great way of incorporating sauerkraut into something a little more familiar, and it's very tasty.

Dish it up as you would with coleslaw. It is a good accompaniment to pulled pork, barbequed meats, or as part of a selection of salads.

SERVES 6–8

¾ cup Sauerkraut (page 16)
⅓ cup kale leaves, thinly sliced
1 cup carrots, peeled and coarsely grated
½ small red onion, very thinly sliced
1 tablespoon thick yogurt
1 tablespoon crème fraîche
1 teaspoon Dijon mustard
pinch of sea salt

Combine all the ingredients in a large mixing bowl and stir well to combine. Serve chilled.

This salad will keep in the fridge, covered with plastic wrap, for up to 3 days.

BIGOS

Known as "Hunter's Stew" in Poland, where Bigos is a national staple, this is a warming and nourishing way of serving sauerkraut. Even more delicious made ahead and served the following day, this is a perfect one-pot dish to feed a crowd.

SERVES 4

1 tablespoon vegetable oil
1 large white onion, thinly sliced
¾ lb pork shoulder, chopped into 1in pieces
1 smoked kabanos sausages, cut into ½ in thick slices
1 oz mixed dried mushrooms, soaked in ½ cup boiling water and drained, reserving the soaking liquid
3 cups cremini mushrooms, cut into 1in thick slices
2¾ cups chicken stock
½ cup heavy cream
Salt and plenty of black pepperr

TO SERVE

¼ cup Sauerkraut (page 16), or more to taste
a handful of chopped parsley

Heat the oil in a large saucepan, wok or casserole dish and add the onion. Cook for about 15 minutes over over medium heat, so that the onion becomes soft and fragrant.

Now, add the pork and cook until lightly browned and slightly caramelized around the edges. Next, add the sausage and mushrooms. Stir together well.

Pour in the mushroom soaking liquid, chicken stock, and cream, season, stir well, and cook over medium heat on the stove for about 45 minutes. The sauce should thicken and reduce by about a third.

Serve hot in bowls, topped with the sauerkraut and parsley on top and some rye sourdough on the side to mop up any juices.

Any leftovers will keep very well in the fridge for up to 3 days.

SAUERKRAUT, BACON, AND POTATO SOUP

This is a lovely way to use sauerkraut. It's a simple recipe, and my take on a soup that is widely enjoyed in the Alsace region of France. It's warming, deeply comforting, and satisfying but also very straightforward to make. It can be ready in under half an hour, with the added bonus of not requiring a blender. I like to add a spoonful of sauerkraut to the soup just as you are serving, so that you get the maximum nutritional benefit from eating it raw.

Serves 4

2 tablespoons olive oil
1 large white onion, thinly sliced
⅔ lb smoked bacon, finely diced
2½ cups vegetable or chicken stock
2 small potatoes, peeled and cut into ½ in cubes
sea salt and freshly ground black pepper
¼ cup Sauerkraut (page 16)

Heat the olive oil in a large saucepan over medium heat. Add the onion and bacon and cook gently until the onion has softened slightly and the bacon is beginning to brown lightly, which should take about 10 minutes.

Pour in the stock, add the potatoes, and season with salt and pepper. Simmer for about 10 minutes until the potato is tender.

Check the seasoning and pour into warmed bowls and top with the sauerkraut, adding more if you like. The soup is excellent served with some sourdough bread to dunk into it.

KIMCHI AND TOFU NOODLE SOUP

Gochujang is a savory chile paste from Korea. Its pungent flavor and red color are quite distinctive and add a richness and depth to this warming, restorative soup.

Serves 4

- 1 quart chicken or vegetable stock
- 6 scallions, thinly sliced at an angle (reserve a couple of slices back for garnishing)
- 1 teaspoon gochujang chile paste (available from Asian supermarkets and online)
- 1¾ cups shiitake mushrooms, thinly sliced
- 1 tablespoon dark soy sauce
- 1 tablespoon sunflower oil
- 1½ cups Kimchi (page 25)
- 1½ cups beansprouts
- 7 oz dried fine egg noodles
- 5 oz silken tofu, cut into ½ in cubes
- 1 teaspoon toasted sesame oil

Pour the stock into a large saucepan and set over medium heat. Add the scallions, chile paste, mushrooms, and soy sauce and bring to a boil. Simmer gently for 5 minutes.

Meanwhile heat the oil in a large frying pan, add the kimchi and beansprouts and stir-fry for 2–3 minutes until heated though.

Cook the noodles according to the instructions on the package and drain into a colander. Divide the noodles between 4 large soup bowls and scatter some of the tofu cubes over each. Top each bowl with a pile of stir-fried kimchi and beansprouts.

Just as you are about to serve, stir the sesame oil into the broth and ladle it over the kimchi. Serve immediately, scattered with the reserved scallion slices.

KIMCHI

Kimchi is an essential component of Korean cuisine, as it is served with almost every meal. It is still made in the fall, in a UNESCO-protected process called Kimjang, when families come together to make their own recipes, which are passed down through the generations. With many regional differences in ingredients and methods, making and eating kimchi is a firm part of Korean heritage. My recipe is for a slightly sweet, tangy kimchi with a crunchy texture. I prefer to thinly slice the cabbage, but you could chop it into chunky pieces if you wish. Personally, I like everything cut up quite small.

MAKES 1 × 1.5-QUART JAR

1¾ lbs total weight of organic white cabbage, thinly sliced, and Chinese-leaf cabbage, cut into 2 in chunks, using more or less of each, as you prefer.
½ cup fresh ginger, peeled and finely grated
6 garlic cloves, peeled and finely chopped
½ cup fresh red chiles, such as fresno or serenade, thinly sliced (leaving the seeds in)
3 organic carrots, peeled and coarsely grated
1 bunch of organic scallions, thinly sliced
1⅔ cups fish sauce
⅓ cup palm sugar
zest and juice of 2 limes
¾ cup filtered water (see page 8)

you will need a 1.5-quart glass Le-Parfait-style jar with a rubber seal, sterilized according to the instructions on page 11

Place the cabbage, Chinese leaf, ginger, garlic, chiles, carrots, and scallions in a large mixing bowl and mix well together with your hands until evenly combined. Transfer the mixture to a 1.5-quart jar.

Add the fish sauce, sugar, lime zest and juice, and water to a pitcher and stir to dissolve the palm sugar. Pour into the jar, stir well with a wooden spoon or spatula, and press down any vegetables that are poking out of the liquid. Close the lid and set aside to ferment on the kitchen countertop for at least a week. When the kimchi is ready it should smell strongly of its component ingredients, but not unpleasant. It won't change drastically in appearance, but the vegetables will soften a little.

The kimchi keeps for up to 2 months in a cool, dark place. Once opened, store in the fridge and eat within a month.

TIP:

Ensure the vegetables are submerged in the brine at all times to inhibit mold from forming on the surface.

KOREAN TOFU STIR-FRY WITH KIMCHI

This stir-fry is my version of a classic Korean dish, known as Japchae. It uses sweet potato noodles, which are an important staple in the Korean diet, and which are gluten-free. Often called glass noodles, because they become clear when cooked, they are quite neutral in flavor and have a sticky texture.

Serves 4

7 oz marinated tofu, cut into ½ in cubes
1¾ oz sweet potato noodles, or vermicelli (available from Asian supermarkets and online)
1½ tablespoons sunflower oil
6⅔ cups baby spinach leaves, stalks removed
¼ cup dark soy sauce
1 tablespoon sesame oil
1 tablespoon superfine or brown sugar
1 bunch of scallions, thinly sliced
2⅓ cups shiitake or cremini mushrooms, thinly sliced
4 garlic cloves, peeled and crushed
1 small carrot, peeled and cut into thin matchsticks
1 zucchini, cut into thin matchsticks

To serve

toasted sesame seeds
¼ cup Kimchi (page 25)

Preheat the oven to 400°F and line a baking sheet with nonstick parchment paper. Arrange the tofu cubes in a single layer on the baking sheet and bake for 20–30 minutes or until golden, firm, and crisp around the edges.

Meanwhile, cook the noodles according to the instructions on the package. Drain and set aside.

Place a large wok over high heat. Add the sunflower oil and allow it to heat for a minute. Put in all of the remaining ingredients, except for the noodles, and stir-fry for 3–4 minutes. Add the noodles and continue to stir-fry for an additional 2 minutes or until they are heated through.

Serve the stir-fry in large bowls, topped with the baked tofu. Finish with a sprinkling of sesame seeds and a generous spoonful of kimchi on top.

KIMCHI PANCAKES

These pancakes, made by incorporating kimchi into a thick batter mixture, are extremely popular in Korea, where they are known as Kimchi Jeon. They can also be made with whole wheat flour, or half whole wheat and half white flour, depending on your preference.

SERVES 4 (MAKES 4 LARGE PANCAKES)

1¼ cups all-purpose flour
pinch of sea salt
1 large egg, beaten
1 cup water
1 cup Kimchi (page 25), drained
1 large carrot, peeled and coarsely grated
a little sunflower oil, for cooking

Sift the flour into a mixing bowl and stir in the salt. Make a well in the center, add the beaten egg and water and whisk the ingredients together to form a thick batter. Finally, stir in the kimchi and grated carrot.

Heat an 8 in-diameter frying pan over high heat. Add a couple of drops of oil and swirl it around to coat the base of the pan.

Pour in roughly a quarter of the batter and swirl the pan so it spreads out in an even layer; it will be quite thick. Cook for about 3 minutes on each side until golden brown and cooked through in the center (this bit takes the longest). After you have flipped the pancake over, it helps if you press down on it using a spatula to ensure the center is cooked through. Once the pancakes are cooked, keep them warm on a baking tray lined with nonstick parchment paper in a low oven (220°F) while you cook the rest. Serve as an appetizer, cut into triangular slices, with a cold beer.

KOREAN PORK WITH KIMCHI

Spicy marinated pork is a popular dish served all over Korea. Accompanied by a generous side of kimchi and rice, it makes an easy but scrumptious meal. This dish contains gochujang, which is a hot and fiery Korean chile paste. If you prefer your food less hot, you can leave it out. The sauce will be much darker in color, but it will still taste wonderful.

Serves 4

For the pork

- ¾ lb boneless pork shoulder, cut into ½ in thick strips
- 2 tablespoons vegetable or peanut oil
- 2 red bell peppers, thinly sliced (seeds and core discarded)
- 5 scallions, thinly sliced on an angle—reserve some to finish

For the marinade

- ¼ cup dark soy sauce
- ¼ cup gochujang chile paste (available from Asian supermarkets and online—optional)
- 2 tablespoons Shaoxing rice wine
- 6 tablespoons dark brown sugar
- 2 tablespoons toasted sesame oil
- 4 garlic cloves, peeled and crushed
- 2 tablespoons peeled and finely grated fresh ginger
- 2 teaspoons mild chile powder
- ¼ teaspoon crushed black peppercorns

To serve

- 2⅓ cups sticky rice
- ¼ cup Kimchi (page 25)
- ¼ cup toasted sesame seeds
- 2 scallions, thinly sliced on an angle

Mix all the ingredients for the marinade together in a large mixing bowl and add the pork. Cover the bowl with plastic wrap and set aside to marinate in the fridge for at least an hour or preferably overnight.

Heat 1 tablespoon oil in a large frying pan or wok over high heat until smoking hot. Add the pork, reserving the marinade, and stir-fry for 15 minutes until cooked through and nicely caramelized. Remove the pork from the pan, and set aside on a warm plate while you stir-fry the vegetables.

Return the wok to high heat with the remaining oil. Add the bell peppers and scallions and stir-fry for 2–3 minutes. Return the pork to the pan with the marinade and stir-fry briefly to combine everything and heat through.

To serve, spoon the stir-fry on to a large plate and sprinkle over the toasted sesame seeds and scallions. Accompany with the kimchi and some sticky rice.

KRAUT-CHI

Kraut-chi is a popular hybrid of German sauerkraut and Korean kimchi. It sounds a bit odd, but it is a lovely blend of flavors and textures and is extremely versatile. It makes an excellent side served with salads, omelets, or all manner of spicy dishes.

MAKES 1 × 1-QUART JAR

- 4⅓ cups organic white cabbage, very thinly sliced
- 1 bunch of organic scallions, thinly sliced
- 2 organic carrots, peeled and coarsely grated
- 1 in piece of fresh ginger, peeled and finely grated
- 4 teaspoons sea salt
- 1 red chile, finely chopped
- 2 tablespoons Water Kefir (page 126)

you will need a 1-quart glass Le-Parfait-style jar with a rubber seal, sterilized according to the instructions on page 11

Place all the ingredients in a large ceramic or glass mixing bowl and toss together with your hands until all the ingredients are well combined. Pack into a 1-quart jar, pressing down well to pack the vegetables in. Close the lid.

Set aside on the kitchen countertop for 5 days. After this time the kraut-chi will smell lightly vinegary and the vegetables will have softened a little.

The kraut-chi will keep for up to 2 months. Once opened, store in the fridge and eat within a month.

FERMENTED VEGETABLES THROUGH THE SEASONS

Makes 1 × 1-quart jar

Spring—Carrots with dill

5 cups organic carrots, peeled and cut into ½ in thick sticks
Large sprig of dill

Summer—Fennel and radish with dill

5 cups organic fennel, cut from top to bottom into ⅛ in thick slices
1¾ cups organic radishes, halved lengthwise
Large sprig of dill

Fall—Spiced cauliflower

5 cups organic cauliflower florets
1 tablespoon curry powder
1 tablespoon chili powder
1 teaspoon cayenne pepper
1 teaspoon ground turmeric
3 garlic cloves, peeled and crushed

Winter—Beet, apple and ginger

3 cups organic beets, peeled and cut into slices, ½ in thick
1⅔ cups organic apples, cored and cut into slices, ½ in thick
⅔ cup organic fresh ginger, peeled and cut into slices, ⅛ in thick

For the ferment

2 cups filtered water (see page 8)
¼ cup Water Kefir (page 126)
2 tablespoons sea salt

you will need a 1-quart glass Le-Parfait-style jar with a rubber seal, sterilized according to the instructions on page 11

This is a very simple method for fermenting fresh vegetables and is a great way of getting started with fermentation. The vegetables will be ready to eat in under a week and retain a delicious crunch. I actually think fermenting them in this way enhances their flavor.

You can use this basic method to ferment a whole range of different vegetables. Simply substitute all or some of the vegetables for your own choice of veg.

Pack the prepared vegetables (and fruit, if using) into a 1-quart glass jar, layering them up if you are using different types. Put in the sprig of dill whole (if using), fitting it down the side of the jar.

To make the ferment, pour the filtered water into a large measuring cup and stir in the water kefir and salt.

Pour the liquid ferment over the vegetables, making sure that all of the vegetables are fully submerged. Fasten the lid.

Set aside on your kitchen countertop for 5 days. When the vegetables are ready to eat they will be slightly softened. The color from the beet and radishes does bleed. You can enjoy your vegetables as they are. Just fish out what you want to eat from the jar, and reseal the lid. They're best eaten within a month of being made.

Once opened, store in the fridge and consume within 2 weeks.

Tips:

If substituting with other vegetables of your choice, ensure that all root vegetables are peeled and slice your vegetables to ½ in thickness.

Only use firm-textured vegetables in this way, such as green beans, broccoli, or parsnips, for example. Soft vegetables, such as eggplant, zucchini, or tomatoes won't work.

RUMTOPF

This makes a scrumptious dessert, served with cream or ice cream. It is very fruity and very boozy; you need only a small quantity at a time. Since it keeps well, it makes a great present.

MAKES 1 × 1-QUART JAR

enough organic soft fruit to fill your jar (approximately 5 cups)—a mixture of strawberries, apricots, cherries, blackberries, and red- and blackcurrants is delicious
2½ cups organic cane sugar
11 fl oz dark rum

you will need a 1-quart glass Le-Parfait-style jar with a rubber seal, sterilized according to the instructions on page 11

Pick over the fruit and remove any stalks or pits. Give the fruit a quick rinse under the tap and halve or slice any large fruit so everything is reasonably uniform in size.

Layer the fruit inside the jar, taking care not to bruise it. Pour in the sugar and rum, making sure that all of the fruit is fully submerged. Fasten the lid and shake the jar very gently to distribute the sugar.

Set aside on the kitchen countertop or in a cupboard for at least 6 weeks before eating. It should be even better after 2–3 months, by which time the fruit will be even softer, and boozier. The rumtopf will smell sweet and alcoholic.

Once opened, store in the fridge and consume within 2 weeks.

FERMENTED BERRIES AND CHERRIES

This is a quick way to ferment any mixture of berries and cherries you like. By fermenting the berries, you are adding to their nutritional value because the beneficial microorganisms involved in the process of fermentation add enzymes, protein and B vitamins to raw ingredients. Fermented fruit is delicious added to smoothies or eaten on its own, accompanied by some homemade yogurt (see page 44). Fermenting fresh fruit is a good way of using up gluts of fruit from the garden or market because fermenting fruit extends its shelf life. Make sure your fruit is fresh and in good condition (no bruises) before you start fermenting.

MAKES 1 × 1-QUART JAR

enough organic berries and cherries to fill your jar (about 5 cups)—most combinations work well here, my favorite one is raspberries, blackberries, strawberries, blueberries, and cherries
6 tablespoons Water Kefir (page 126)
½ cup superfine sugar
pinch of sea salt

you will need a 1-quart glass Le-Parfait-style jar with a rubber seal, sterilized according to the instructions on page 11

TIPS:

Most combinations work well here. Just make sure that the fruit is in perfect condition to inhibit the formation of any mold.

You could halve the quantities if you wish, but do fill your jar to the brim with fruit and keep it submerged at all times otherwise mold could start to form on the surface.

Pick over the fruit and remove any stalks and pits, including the cherry pits. Discard any fruit that is bruised or damaged. Give the fruit a quick rinse under the tap. It doesn't matter if the fruit are of different sizes—it will give a nice texture to the end result.

Add the fruit into the jar, gently, and avoid packing it down too tightly or it will bruise and squash. Pour in the water, sprinkle in the sugar and salt and press the fruit down gently so that it is completely submerged.

Fasten the lid and set aside on the kitchen countertop to ferment for 24–48 hours. The color from the fruit will bleed, so it will look quite dark in color. The fruit will softe, and some of the berries may burst or break up a little. Once fermented, transfer the fruit to the fridge where it is best enjoyed within a week

FERMENTED FRUIT COMPOTE

This compote makes a really good dessert, accompanied by some vanilla ice cream. It's a great way to make use of any seasonal gluts of fruit. Most soft fruit works well here, however my favorite combination is raspberries, blackberries, red- and black currants, and apricots.

Makes 1 × 1-quart jar

enough organic soft fruit to fill your jar (about 5 cups)—any mixture of berries or soft fruit is delicious
1 cup + 2 tablespoons superfine sugar
7 fl oz brandy

you will need a 1-quart glass Le-Parfait-style jar with a rubber seal, sterilized according to the instructions on page 11

Pick over the fruit and remove any stalks or pits. Give the fruit a quick rinse under the tap and halve or slice any large fruit so everything is roughly the same size.

Layer up the fruit inside the jar, taking care not to bruise it. Pour in the sugar and brandy and stir gently to distribute the sugar.

Fasten the lid and set aside to ferment on the kitchen countertop for 7 days, stirring every 2–3 days. Once ready, the compote will look quite dark in color and will smell of fruity brandy.

This one is best eaten within a month. Once opened, store in the fridge and consume within a week.

YOGURT & LABNEH

HOMEMADE WHOLE-MILK YOGURT

Making your own yogurt is very straightforward and tremendously satisfying. It is not necessarily cheaper than buying it at the store, but it is a lovely thing to do—especially if you can source some good, local milk. The milk powder thickens the yogurt, and I think it tastes better with it included. If you prefer a creamier yogurt still, you can substitute ½ cup of the milk with half and half.

Makes approx. 2 cups

2 heaped tablespoons organic live yogurt
2 cups organic whole milk
3 tablespoons milk powder

you will need a digital thermometer and a 1-quart insulated flask, spotlessly clean

Tip:

It's vital that everything is spotlessly clean when making yogurt. To ensure your equipment is in perfect condition, place the whisk and metal spoon inside the mixing bowl and fill it to the brim with boiling water before use. Dip the thermometer tip in, too. Pour the water away, dry the equipment using a clean dish towel or paper towels before using.

Place the yogurt in a spotlessly clean large glass or ceramic mixing bowl and allow it to come up to room temperature, approx. 20–30 minutes.

Pour the milk into a clean saucepan and heat very gently over very low heat until it reaches exactly 115°F—don't let it get any warmer than this or it will kill the live cultures in the yogurt when the two are combined. Remove the pan from the heat, sprinkle the milk powder over the surface and whisk it in thoroughly. Carefully pour the warm milk over the yogurt in the bowl and stir well with a metal spoon.

Pour the yogurt into the flask. Screw on the lid and set aside on the kitchen countertop overnight.

By morning, your milk should have thickened and turned into yogurt. Decant it into jars or a plastic container, cover with plastic wrap or a lid, and store in the fridge. Eat within 5 days.

STRAWBERRY FROZEN YOGURT

Bright pink and intensely fruity, this heavenly frozen yogurt is a great way of making use of sweet, ripe berries at their best. This is also excellent made with homemade coconut yogurt (see opposite page) instead of homemade whole-milk yogurt for a vegan dessert. Even better, you don't need an ice cream maker to make this recipe.

MAKES APPROX. 1-QUART

3 cups strawberries, rinsed and hulled
⅔ cup superfine sugar
1 cup Homemade Whole-milk Yogurt (page 44)
1 teaspoon fresh lemon juice

you will need a blender or food processor and a 1-quart plastic container with a lid

Slice the strawberries into quarters. Transfer to a mixing bowl, sprinkle in the sugar, and stir with a metal spoon until nicely combined. Cover the bowl loosely with plastic wrap and set aside at room temperature for 1 hour to allow the fruit to macerate.

Add the strawberries and all of the juices into a blender or food processor. Add the yogurt and fresh lemon juice and blitz until smooth. If you wish, you can strain the mixture through a nylon sieve at this stage to remove the seeds for a totally smooth finish, but this isn't essential.

Pour the mixture into a plastic container with a lid and transfer it to the freezer for 1 hour. Then remove the mixture from the freezer, put into a stand mixer bowl or large mixing bowl and whisk vigorously to break up all the ice crystals. Transfer the ice cream back to the plastic container, put the lid back on, and return to the freezer. Repeat this process three times, setting your timer at 1-hour intervals during which time the ice cream will become smooth as it sets.

After you have whisked the ice cream for the fourth time, return it to the freezer for a few more hours until it has frozen solid. Leave it there until you are ready to eat. Enjoy within 2 months.

HOMEMADE COCONUT MILK YOGURT

Everyone seems to love coconut yogurt. It's so popular (and expensive), I wanted to have a go at making it myself. Most commercial brands are not fermented and are thickened using tapioca starch flour, but I prefer to make mine using agar agar, which is actually a form of dried seaweed and is completely natural, rich in minerals, and vegan, too. This yogurt is thinner than store-bought versions, but still as delicious, and can be used to make the frozen yogurt and ice cream recipes in the book.

Makes approx. 14 oz

1 × 14 oz can of full-fat coconut milk
2 tablespoons agar agar flakes

you will need a digital thermometer

Pour the coconut milk into a small saucepan and warm over very low heat until it reaches exactly 115°F. Remove the pan from the heat, sprinkle the agar agar flakes over the surface, and whisk in thoroughly. Continue to whisk over medium heat for 5–10 minutes until the agar agar flakes are dissolved. Let cool.

After 30 minutes, whisk again vigorously to bring the yogurt to a smooth, even texture, as lumps can form as it cools. An electric hand whisk or food processor is ideal to use here. Pour the mixture into a clean earthenware or glass bowl.

Cover the bowl with plastic wrap, allow to cool, and refrigerate overnight.

Once the yogurt has set, spoon it into a clean glass, plastic, or ceramic jar, put on the lid or cover with plastic wrap and transfer to the fridge. Consume within 5 days.

COCONUT YOGURT ICE CREAM

This is the next step once you've perfected making your own coconut yogurt— making it into ice cream! You don't need an ice cream maker to make ice cream from scratch. You do need to be in for a couple of hours though, as the ice cream needs whisking a few times as it freezes to prevent ice crystals from forming. This ice cream is fantastic served on its own, especially after spicy food, and is also lovely with fresh fruit, or a squeeze of fresh lime and some coconut flakes.

Makes approx. 1-quart

1 × 14 oz can of full-fat coconut milk
¾ cup superfine sugar
14 oz Homemade Coconut Yogurt (page 47)

you will need a 1-quart plastic container with a lid and a set of electric beaters or stand mixer are useful to use, if you have them

Tip:

Greek yogurt also works well in this recipe instead of the homemade coconut yogurt, producing a slightly sharper flavor.

Pour the coconut milk into a large saucepan and add the superfine sugar. Set the pan over medium heat and keep stirring with a wooden spoon until the sugar has dissolved. Don't let it come to a boil. Once the sugar has dissolved, remove the pan from the heat and set aside to cool to room temperature before whisking in the yogurt.

Pour the mixture into a plastic container with a lid and transfer it to the freezer for 1 hour. Then remove the mixture from the freezer, pour into a stand mixer bowl or large mixing bowl and whisk vigorously to break up all the ice crystals. Transfer the ice cream back to the plastic container, put the lid back on, and return to the freezer. Repeat this process three times, setting your timer at 1-hour intervals during which time the ice cream will become smooth as it sets.

After you have whisked the ice cream for the fourth time, return it to the freezer for a few more hours until it has frozen solid. Leave it there until you are ready to eat. Eat within 2 months.

VANILLA YOGURT PANNA COTTA

This is a very elegant dessert with a creamy, yet tangy flavor. It's superb served with some fresh or poached fruit. There are a number of ways of making panna cotta. You can pour the mixture into ramekins to serve the panna cotta in, or pour the mixture into lightly greased muffin pans and carefully turn them out onto individual dishes. If you are turning out your panna cotta, you'll need to use the extra sheet of gelatin, to make it a little firmer. You can use agar agar flakes too, to make a vegetarian version. Panna cotta made with agar agar is best made and served in individual ramekins because the set is softer than with gelatin.

Serves **6**

peanut or vegetable oil, for greasing
4 sheets of gelatin if turning out, or 3 sheets if using ramekins; alternatively 3 tablespoons agar agar flakes
1 cup heavy cream
¾ cup superfine sugar
1 vanilla pod, split in half lengthwise
1 cup Homemade Whole Milk Yogurt (page 44), or Greek store-bought yogurt

you will need 6 × 3.5 oz ramekins or pretty glasses for serving

Lightly grease the ramekins with oil and set aside. (If you don't have ramekins, you could use some pretty glasses instead—in which case there is no need to grease them.)

Prepare the gelatin according to the instructions on the package.

Pour the cream into a small saucepan and add the sugar and vanilla pod. Set the pan over low heat and warm very gently until the cream just starts to simmer. Stir gently at regular intervals to dissolve the sugar.

Remove the pan from the heat, add the gelatin mixture, and stir with a wooden spoon until well combined. Set aside to cool for 20 minutes or so, until the mixture is lukewarm, and then stir in the yogurt. Be patient: don't add the yogurt too soon, or it will change the texture of the panna cotta.

Divide the mixture between the prepared ramekins (or glasses), discarding the vanilla pod, and set aside until completely cold. Then cover the surface with plastic wrap and transfer to the fridge to set overnight.

To serve, either present them in ramekins or glasses, or turn the panna cottas out of their molds and accompany with a few berries or some poached fruit.

CARDAMOM AND ROSE YOGURT CREAM

This lightly perfumed dessert, which is not too dissimilar to a mousse, makes a light and interesting end to a meal. You'll probably only want a small helping of this, because it's very rich and creamy. It is gorgeous served with some little butter cookies, such as shortbread.

Serves 4–6

1 cup heavy cream
1 cup Homemade Whole-milk Yogurt (page 44)
⅓ cup superfine sugar
seeds from 4 green cardamom pods, crush the pods and carefully pick out the seeds, discarding the husks
2–4 drops of rosewater, depending on the strength
crystallized rose petals (optional), to serve

Pour the heavy cream into a large mixing bowl and whisk until soft peaks form. Add the remaining ingredients, except the rosewater, and fold in carefully using a large metal spoon. Sprinkle over the rosewater, just a couple of drops to begin with, and fold in gently. Then taste the perfumed cream and see if you need any more before adding in the remainder.

Cover the bowl with plastic wrap and set aside in the fridge until needed. For an extra special touch, sprinkle over some crystallized rose petals before serving.

Store in the fridge and eat within 3 days.

MINT AND CUCUMBER RAITA

This makes a delicious, cooling accompaniment to spicy Indian foods. It is really lovely served on basmati rice, and also with tandoori and tikka meats (see page 56).

Serves 4

- 1 cup Homemade Whole Milk Yogurt (page 44)
- 3 in piece of cucumber, peeled or unpeeled and coarsely grated
- 2 heaped teaspoons chopped fresh mint
- ½ teaspoon cumin seeds, toasted
- generous pinch of sea salt
- freshly ground black pepper

Place all of the ingredients in a mixing bowl and stir well to combine. Cover the bowl with plastic wrap and set aside in the fridge until needed. Serve chilled.

The raita will keep for up to 3 days, covered with plastic wrap and stored in the fridge.

Tip:

To toast the cumin seeds, place them in a dry pan, and set over medium heat until fragrant, approx. 1 minute. Watch that they don't burn.

TANDOORI-STYLE CHICKEN WITH YOGURT

This is a delicious Indian meal that is so easy to make from scratch. I highly recommend making it the night before you want to eat it, to allow the flavors to develop fully.

It is sensational served with boiled rice, naan bread or Sourdough Saj (page 111) and Cucumber Raita on page 57.

Serves 4

- 2 lbs skinless chicken pieces (a combination of breast, thigh, and drumsticks is nice)
- 2 generous pinches salt
- 2 tablespoons lemon juice
- 1¼ cups Homemade Whole Milk Yogurt (page 44)
- 1 small onion, peeled and chopped
- 1 clove garlic, peeled and chopped
- 1 in piece of ginger, peeled and finely grated
- ½ small green chile, seeded and chopped
- 2 teaspoons garam masala

Place the chicken in a large baking dish. Using a sharp knife, make some deep slashes in the chicken. Sprinkle over the salt and lemon juice and leave to sit for a few minutes.

Now, put all the remaining ingredients in a food processor and blitz to form a smooth paste. If you don't have a food processor, you can chop all the ingredients finely and stir together. Pour the yogurt over the chicken and rub the marinade into the slashes. Cover and leave in the fridge to marinate for at least 8 hours or overnight. It's a great dish to make in advance either the night before or on the morning of the day you want to eat it.

When you are ready to eat the chicken, preheat the oven to 400°F. Place the chicken pieces on a large baking sheet, discarding any excess marinade. Bake for 20-25 minutes until the chicken is thoroughly cooked.

Serve with freshly cooked rice or sourdough saj, raita, lime wedges, and plenty of fresh cilantro.

PERSIAN EGGPLANT STEW
WITH SAFFRON YOGURT

This quick-to-make stew combines a myriad of delicious Middle Eastern flavors. A complete meal in itself, it is perfect for both a weeknight supper or for easy entertaining. Freekeh is a Middle Eastern green (unripe) wheat grain that's picked whole and roasted over wood fires, to give it a delicious smoky, earthy flavor and to burn off the husks. It's now widely available in supermarkets and health food stores.

Serves 4

For the stew

1 cup freekeh
1 quart vegetable or chicken stock
3 tablespoons olive oil
1 large yellow onion, thinly sliced
2 large eggplants, cut into 1 in cubes
1 teaspoon ground cinnamon
2 teaspoons ground cumin
2 teaspoons ground turmeric
sea salt and black pepper
2 tablespoons tamarind paste
2 tablespoons honey

For the saffron yogurt

pinch of saffron, infused in 1 teaspoon boiling water
¾ cup Homemade Whole-milk Yogurt (page 44)
1 small garlic clove, peeled and crushed
1 tablespoon lemon juice
2 tablespoons olive oil
pinch of sea salt

To serve

¼ cup pomegranate seeds
small handful of fresh mint leaves, roughly chopped

Combine the freekeh and stock in a large casserole and set over high heat without the lid on. Bring to a boil then turn down the heat to medium and simmer gently for 15–20 minutes until the freekeh is cooked al dente.

Meanwhile, heat 1 tablespoon of olive oil in a large frying pan and cook the onions over medium heat until they soften, about 15 minutes. Don't let them brown! Once the freekeh is cooked, add the softened onion, reduce the heat to low and simmer gently for around 5 minutes. Keep the pan you used to cook the onions to one side—no need to wash it because you'll be needing it again!

Meanwhile, combine the eggplant chunks in a large mixing bowl and sprinkle in the cinnamon, cumin, turmeric, salt, and pepper. Toss together with your hands to ensure the eggplant pieces are evenly coated with the spice mixture. Add the remaining 2 tablespoons of olive oil to the frying pan and heat over medium—high heat. Put in the eggplant chunks and cook for about 5 minutes until golden, turning frequently.

Transfer the fried eggplant to the casserole, add the tamarind and honey, and stir to combine. Taste, and season with salt and pepper if necessary. Keep the stew warm over low heat with the lid on while you prepare the yogurt. To make the saffron yogurt, combine all of the ingredients in a mixing bowl and stir to combine.

To serve, spoon into a serving bowl and top with the saffron yogurt. Garnish with some pomegranate seeds and fresh mint leaves—as many or as few as you like. Eat immediately.

This is a really nice dish to serve on a large warmed platter or serving dish, placed centrally on the table so that everyone can help themselves.

YOGURT PANCAKES

Another great use for homemade yogurt is in yogurt pancakes. These fluffy, Scotch-style pancakes are delicious served for breakfast or lunch. They are excellent accompanied by fresh or fermented fruit (see page 40)—sliced banana and honey is always a winner—and they are also delicious with maple syrup and bacon.

MAKES 8–12, DEPENDING ON SIZE

¾ cup self-rising flour
2 tablespoons superfine sugar
pinch of salt
1 large egg, beaten
⅓ cup Homemade Whole-milk Yogurt (page 44)
4 tablespoons whole milk
2 tablespoons sunflower oil, for cooking

Sift the flour into a medium-sized mixing bowl and stir in the superfine sugar and salt.

Put the beaten egg, yogurt, and milk in a measuring cup and whisk together. Pour the wet ingredients over the dry ingredients in the bowl and whisk to form a smooth batter without any lumps. Pour the mixture back into the measuring cup, and set aside for 15 minutes. It's much easier to pour the mixture into the pan from a measuring cup than a bowl.

Heat a little oil in a large nonstick frying pan over medium—high heat. Spoon tablespoons of the mixture into the pan and cook until bubbles appear on the surface and the undersides of the pancakes turn medium brown. Flip the pancakes over with a spatula and cook on the other side until golden brown. Serve with the topping of your choice.

TIP:

If the pancakes aren't being eaten immediately, then before you start, preheat the oven to 200°F, cover a large baking sheet with nonstick parchment paper and transfer the pancakes to the tray in the oven, one at a time, to keep warm until required. They're best eaten within half an hour, though.

SWEET YOGURT SCONES

These sweet scones are exceptionally light, fluffy, and moist. The sugary topping is optional but really delicious. It's a trick I learned from Darina and Rachel Allen at Ballymaloe Cookery School, and it makes the scones even more irresistible.

MAKES 8–12, DEPENDING ON SIZE

2½ cups self-rising flour, plus extra for dusting
pinch of sea salt
1 level teaspoon baking powder
¼ cup superfine sugar
½ cup cold salted butter, cubed
⅓ cup whole milk
1⅓ cups Homemade Whole Milk Yogurt (page 44), full-fat, or store-bought Greek yogurt

FOR THE TOPPING

1 egg, beaten
¼ cup superfine sugar

you will need a 2 in biscuit cutter

Preheat the oven to 400°F and line a large baking sheet with nonstick parchment paper.

Prepare the ingredients for the topping. Pour the beaten egg onto a small plate and sprinkle ¼ cup of superfine sugar over a separate plate. Set both plates aside while you prepare the scones.

Sift the flour, salt, and baking powder into a large mixing bowl and stir in the superfine sugar. Add the cubes of butter and rub into the dry mixture using your fingertips. The finished texture should resemble sand; it is important to ensure there are no large chunks of butter remaining.

Combine the milk and yogurt in a measuring cup and whisk briefly to combine. Pour the wet ingredients over the dry mixture in the bowl and mix lightly together. Don't over-mix or the scones will turn out dry and tough; you just want to incorporate the dry ingredients into the wet. (If using a thicker yogurt, such as Greek style, the mixture will be fairly stiff and will only take a few seconds to mix.)

Once the dry and wet ingredients are combined, place the mixture onto a floured work surface and roll out to a thickness of about 2 in. Cut into circles or squares, as you prefer. (Avoid the temptation of re-rolling the scraps; the dough will be overworked and you'll only end up with tough scones.)

To create the lovely sugary topping, dip the top of each scone first in the beaten egg, and then in the sugar and place topping-side up on the baking sheet.

Bake the scones for 12–15 minutes until they are lightly browned and they feel fairly firm but yielding. Transfer to a wire rack to cool slightly before eating. They are especially good eaten within 4 hours.

Any leftovers are best stored in the freezer, as scones are never the same the following day.

TIP:

Scones are so quick and easy to make, I urge you to resist using an electric mixer or processor. I promise you, they'll be much lighter, fluffier, and tender made by hand.

LEMON AND RASPBERRY YOGURT LOAF CAKE

Often described as "not too sweet," this moist, fruity cake is equally delicious served with a steaming hot cup of tea in the afternoon as it is with a spoonful of cream for dessert.

Makes 1 × 2lb loaf cake
Serves 8–10

2 cups all-purpose flour
2 level teaspoons baking powder
pinch of fine sea salt
1⅓ cups superfine sugar
2 tablespoons ground almonds
½ cup softened salted butter
zest of 2 unwaxed lemons
½ cup Homemade Whole-milk Yogurt (page 44), or Greek store-bought yogurt
2 large eggs, beaten
1⅔ cups fresh raspberries (or 1 cup frozen raspberries, defrosted on paper towels)
3 tablespoons demerara sugar

you will need 1 × 2 lb loaf pan

Preheat the oven to 400°F and line your pan with nonstick parchment paper.

Sift the flour, baking powder, and salt into a large mixing bowl. Add the superfine sugar, ground almonds, and butter and rub in with your fingertips until the mixture resembles breadcrumbs.

Add the lemon zest, pour in the yogurt and beaten eggs, and stir gently to combine. Finally fold in the raspberries very gently so as not to bruise them. Spoon the mixture into the prepared pan, level the surface, and sprinkle the demerara sugar evenly over the top.

Bake for 45 minutes—1 hour or until the cake is well-risen and golden brown, and a toothpick inserted into the center of the cake comes out clean. Remove the cake from the oven and set aside to cool in the pan for 15 minutes. Remove from the pan, peel off the parchment paper and place the cake on a wire rack to cool fully.

Best eaten fresh, but will keep in an airtight container for up to 5 days.

SAVORY LABNEH

Labneh is a staple of the Middle East, where it is widely made. It's a dripped yogurt, made by suspending yogurt in a cheesecloth bag to allow the whey to drip out and leave a thick and creamy yogurt, which is called labneh. This version can be enjoyed on its own or with a selection of mezze, and it is often eaten for breakfast with raw vegetables and flatbreads. It is salted for use in savory dishes, and is the key component in the following recipes (see pages 66 or 79).

MAKES APPROX. 1¼ LBS

1 × 2 lb tub of organic wholemilk, Greek or Homemade Whole Milk Yogurt (page 44)
1 tablespoon sea salt

you will need a clean cheesecloth (see page 11) and some string

TIP:

Do keep the whey. It can be used in a number of recipes in the book, including the Sourdough (page 104) or Rye Sourdough (page 109). It is also good added to water to make up the required quantity of liquid needed for pizza dough.

Place the yogurt in a mixing bowl and beat in the salt.

Set a sieve over a large mixing bowl. Drape a large cheesecloth over the surface, so that the base of the sieve is fully lined and the sides of the cheesecloth hang down over the sides of the bowl.

Spoon the yogurt into the center of the cheesecloth, gather up the sides, and tie them tightly with string just above the yogurt. Leave the lengths of string long to tie it up.

Suspend the cheesecloth bag over the mixing bowl by tying it in a position where it can hang and the whey can drip into the bowl. I do this at home by tying the cheesecloth bag onto a cupboard handle over the countertop. Some people like to tie it to a faucet and sit the bowl in the sink below. Set aside in a cool place overnight, during which time the whey will drip through the cheesecloth.

The contents of the cheesecloth bag is your labneh. In the morning, pour the labneh into a clean earthenware or glass bowl, cover with plastic wrap, and set aside in the fridge until needed. (It is helpful to weigh how much labneh you have made at this point so that you can make up any recipes accordingly.) Don't discard the whey (see Tip).

The labneh will keep well in the fridge for up to 3 days.

SPICY LABNEH

In the Middle East labneh is often eaten flavored, and this is a delicious variation on plain labneh, with a little spice and heat; for an extra kick.

MAKES APPROX. 1¼ LBS

1 × 2lb tub of organic whole-milk, Greek, or Homemade Whole-milk Yogurt (page 44)
1 garlic clove, peeled and crushed
1 teaspoon sea salt
2 teaspoons toasted cumin seeds
pinch of chile flakes
small handful of fresh soft herbs, such as parsley, mint, cilantro, or dill, finely chopped
1 tablespoon extra virgin olive oil
you will need a clean cheesecloth (see page 11) and some string

TIP:

To toast the cumin seeds, place them in a dry frying pan, set over low heat and shake the pan from side to side until they start to release their aroma. Pour onto a serving plate as soon as they start to color and set aside to cool.

Combine all the ingredients in a large mixing bowl and beat together well. Set a sieve over a large mixing bowl. Drape a large cheesecloth over the surface, so that the base of the sieve is fully lined and the sides of the cheesecloth hang down over the sides of the bowl.

Spoon the yogurt into the center of the cheesecloth, gather up the sides, and tie them tightly with string just above the yogurt. Leave the lengths of string long to tie it up.

Suspend the cheesecloth bag over the mixing bowl by tying it in a position where it can hang and the whey can drip into the bowl. I do this at home by tying the cheesecloth bag onto a cupboard handle over the countertop. Some people like to tie it to a faucet and sit the bowl in the sink below. Set aside in a cool place overnight, during which time the whey will drip through the cheesecloth.

The contents of the cheesecloth bag is your labneh. In the morning, pour the labneh into a clean earthenware or glass bowl, cover with plastic wrap, and set aside in the fridge until needed. (It is helpful to weigh how much labneh you have made at this point so that you can make up any recipes accordingly.)

The labneh will keep well in the fridge for up to 3 days.

SWEET LABNEH

This is a slightly different version of traditional labneh, which is traditionally made with salt to enhance its savory flavor. By omitting the salt, this sweet labneh can be used in any sweet dishes or just enjoyed on its own.

Makes approx. 1¾ lbs

1 × 2lb tub of organic whole-milk yogurt—Greek yogurt is especially good as it gives a very thick, creamy result

you will need a clean cheesecloth (see page 11) and some string

Set a sieve over a large mixing bowl. Drape a large cheesecloth over the surface, so that the base of the sieve is fully lined and the sides of the cheesecloth hang down over the sides of the bowl.

Spoon the yogurt into the center of the cheesecloth, gather up the sides, and tie them tightly with string just above the yogurt. Leave the lengths of string long to tie it up.

Suspend the cheesecloth bag over the mixing bowl by tying it in a position where it can hang and the whey can drip into the bowl. I do this at home by tying the cheesecloth bag onto a cupboard handle over the countertop. Some people like to tie it to a faucet and sit the bowl in the sink below. Set aside in a cool place overnight, during which time the whey will drip through the cheesecloth.

The contents of the chesecloth bag is your labneh. In the morning, put the labneh into a clean earthenware or glass bowl, cover with plastic wrap, and set aside in the fridge until needed. (It is helpful to weigh how much labneh you have made at this point so that you can make up any recipes accordingly.)

Store in the fridge and use within 3 days.

Tip:

You can use any whole-milk yogurt for this sweet version, or homemade, too. Avoid using low-fat yogurt, as the resulting consistency will be too thin.

PRESERVED LABNEH WITH CHILE, LEMON, THYME AND OLIVE OIL

Labneh can be preserved for up to 3 months by preserving it in oil, but it must be completely submerged at all times. Rolling labneh into balls and coating them with an aromatic combination of flavors is not only delicious, but a great way to make your labneh last. What's particularly nice is that you can scoop out as many or as few balls as you like, as and when you want to use them. They're gorgeous served in a bowl as part of a mezze selection, or spread on a flatbread, such as the saj on page 111.

Makes 1 × 1-quart jar
Serves 6

1¼ lbs Savory Labneh, drained weight (page 64)
6 heaped teaspoons chile flakes
6 heaped teaspoons chopped thyme leaves
zest of 4 unwaxed lemons, finely grated (ideally using a microplane grater)
approx. 3½ cups extra virgin olive oil

you will need a 1-quart glass Le-Parfait-style jar with a rubber seal, sterilized according to the instructions on page 11

Put the labneh in a large mixing bowl. Combine the chile flakes, thyme, and lemon zest on a large plate.

Pinch off a little piece of labneh about the size of a walnut and shape it into a ball. Roll it in the herb and spice mixture so it is well coated and place in the bottom of the jar. Repeat with the rest of the mixture.

Add extra virgin olive oil until the jar is full, ensuring that all of the labneh balls are completely submerged, and fasten the lid.

The labneh balls can be eaten immediately. Otherwise,transfer the jar to a cool, dark place, where the labneh balls will keep for up to 3 months. They can be kept in the fridge, but the olive oil will solidify.

Once opened, use within 2 weeks.

Tip:

Once you have used up all of the labneh balls you can use the flavored oil as a salad dressing or for cooking, so it won't go to waste.

ZUCCHINI, LABNEH, AND THYME TART

This tart is incredibly easy to make and looks so elegant. It's delicious served with a salad as a main meal, or you could wrap a slice in foil to take out with you as a packed lunch.

Serves 4–6

- 1 × 13 oz pack of ready-made, ready-rolled puff pastry
- ¼ cup Savory Labneh (page 64)
- 1 tablespoon milk or 1 egg, beaten
- 2 large or 6 baby zucchini, ends trimmed and sliced at an angle, approx. 2 in thick
- 3 tablespoons Parmesan, finely grated
- 1 tablespoon thyme leaves
- sea salt and black pepper

Preheat the oven to 350°F.

Unroll the puff pastry onto a large baking sheet. Take a sharp knife and score a border around the edge of the pastry, running roughly 1 in parallel to the sides.

Spread the labneh over the middle of the pastry, taking it up to, but not over, the pastry border. Brush the border with a little milk or beaten egg, and season the labneh all over with salt and pepper.

Now, arrange the zucchini slices over the labneh, ensuring the labneh is covered. Sprinkle the Parmesan evenly over the surface and scatter over some thyme leaves.

Bake for 18–22 minutes, by which time the border should have puffed up nicely and the topping will be golden brown. Serve immediately while still hot, or it is good cold the following day.

LABNEH WITH FIGS, HONEY AND WALNUTS

This is a gorgeous dessert, or even a decadent breakfast, that is assembled in minutes. Incorporating some of the best flavors of the Mediterranean, it makes the perfect easy summer dessert, preferably eaten outdoors. Present a large bowl of this to the table, and you'll be very popular indeed.

This recipe is also lovely with slices of ripe, juicy pear instead of figs.

Serves 6

approx. 1 lb Sweet Labneh, drained weight (page 67)
9–10 teaspoons good quality honey
5 teaspoons chopped walnuts
6 figs, quartered

Prepare the labneh as on page 67 and put into a serving bowl. Add the honey and beat well with a wooden spoon to combine—if using runny honey you could swirl it in if you prefer. Sprinkle the walnuts over the top and decorate with the quartered figs. Serve immediately, drizzled with a little extra honey, if you like.

LABNEH WITH POMEGRANATE, MINT AND SUMAC

This is another lovely, fresh tasting dessert, this time drawing on flavors from the Middle East. Sumac adds a lemony tang and is used widely in a variety of Mediterranean and Middle Eastern dishes, where it is added to both sweet and savory dishes.

SERVES 6

1 lb Sweet Labneh, drained weight (page 67)
3 heaped tablespoons superfine sugar
seeds from 1 large pomegranate
1½ teaspoons ground sumac
1 tablespoon roughly chopped mint leaves

Prepare the labneh as on page 67, place into a serving bowl and beat in the sugar. Sprinkle over the pomegranate seeds and sumac at the last moment and finish with a scattering of mint leaves. Serve immediately.

This recipe is also lovely with pomegranate, sliced mango, and mint.

CHOCOLATE LABNEH CHEESECAKE

This recipe is for a dark chocolate cheesecake. If you prefer your cheesecake a bit sweeter, you can substitute the dark chocolate for milk chocolate and use milk chocolate cookies for the base.

Serves 8–10

For the base

6½ oz (approx. 12) dark chocolate graham crackers
6 tablespoons salted butter, melted

For the filling

7 oz dark chocolate (approx. 70 percent cocoa solids)
1¼ cups heavy cream
½ lb sweet Labneh, drained weight (page 67)
1 cup superfine sugar

For the topping

3½ oz dark chocolate (approx. 70 percent cocoa solids), grated, chopped or curled

you will need a 10 in round springform cake pan

Line the base and sides of the cake pan with nonstick parchment paper.

Break the chocolate for the filling into pieces and place in a heatproof bowl over a pan of gently simmering water. Once the chocolate has fully melted, set it aside to cool until needed.

To make the base, put the chocolate graham crackers in a food processor and process to coarse crumbs. (If you don't have a food processor, you can put the crackers into a plastic bag and crush them with a rolling pin.) Add the cracker crumbs into a mixing bowl and stir in the melted butter. Spoon the mixture into the prepared pan and flatten it down evenly to cover the base. Set aside.

To make the filling, pour the cream into a mixing bowl and whip until soft peaks form. Place the labneh in a separate mixing bowl and beat well with a wooden spoon to soften the mixture. Add the whipped cream and superfine sugar to the labneh and fold in gently. Once the melted chocolate has cooled, carefully fold it into the labneh cream. Spoon the filling over the graham cracker base and smooth the surface with a palette knife.

Cover the whole thing loosely with foil and set aside to firm up in the refrigerator for at least 4 hours or preferably overnight.

Just before serving, sprinkle the grated chocolate over the surface of the cheesecake and cut into wedges.

SUPER-HEALTHY LABNEH CHEESECAKE WITH HONEY

As cheesecakes go, this one is really rather virtuous. It's made using natural, nutritious ingredients, and is free from wheat, gluten, and refined sugar. Using labneh in the cheesecake mixture gives a slightly more acidic finish, not too dissimilar from a sharp cream cheese.

Serves 12

For the base

2½ cups nuts of your choice—such as cashew nuts, or a mixture of cashews, almonds, and walnuts
5 oz pitted dates
⅓ cup coconut oil
a few drops of vanilla extract

For the filling

1 lb Sweet Labneh, drained weight (page 67)
1 teaspoon vanilla extract
¼ cup raw honey, plus extra to drizzle

you will need a 10 in round springform cake pan and a food processor

Line the base and sides of the cake pan with nonstick parchment paper.

To make the cheesecake base, place the nuts and dates in a food processor and process until the nuts form small chunks. Add the coconut oil and vanilla extract and pulse briefly again, just until the mixture starts to bind together. Add the mixture into the prepared pan and press down well with the flat of your hand to form an even layer. Cover with plastic wrap and transfer to the refrigerator to firm up while you make the cheesecake filling.

To make the filling, combine the labneh, vanilla extract, and honey in a large mixing bowl and beat together until evenly blended. Spread the mixture over the base and level the surface with a palette knife.

Cover the surface loosely with foil and transfer to the refrigeratorto set for a minimum of 8 hours.

When you are ready to serve, remove the pan and transfer the cheesecake to a serving plate. Drizzle some honey over the top and cut into slices.

This is a lovely dessert to make for a gathering. It's best eaten within three days.

SAJ WITH LABNEH AND ZA'ATAR

This is one of the most memorable things I ate on my first trip to the Middle East. I sat on a shady terrace, escaping from the blistering sun, and devoured this for lunch. I suggest you roll the bread up, with the labneh spread inside, a bit like a wrap, but you could always cut the saj into triangles and serve them alongside little bowls of the labneh and za'atar for dipping.

Serves 1

- 2 tablespoons Savory Labneh (page 64)
- 1 warm Sourdough Saj flatbread (page 111)
- 2 teaspoons Za'atar (see below)

Spread the labneh over the bread, sprinkle over the za'atar and roll up to serve as a wrap. Alternatively you can serve the individual components separately, as above.

HOMEMADE ZA'ATAR

This delicious blend of herbs and spices is widely used in Middle Eastern cookery. Homemade za'atar tastes so much better than store bought. It takes moments to make and is a really useful flavoring to have on standby to sprinkle over dips, as a rub for meat and vegetables, or to spread over dough to make delicious breads.

Makes 1 × 3.5 oz jar

- 4 teaspoons sesame seeds
- 8 teaspoons dried oregano
- 4 teaspoons dried marjoram
- 4 teaspoons ground sumac
- 1 teaspoon sea salt
- 4 teaspoons cumin seeds, toasted and ground

you will need a 3.5 oz clip-top jar with a rubber seal lid

Spoon all of the ingredients into a small jar and shake together to combine.

Store for up to 6 months in a sealed airtight container.

SHRIKHAND

This gorgeous Indian sweet, made from strained yogurt (labneh), is widely eaten across the regions of Kerala, Gujarat and Maharashtra in India. It is a beautifully fragrant, delicately flavored dish that makes a perfect ending to an Indian meal.

SERVES 4

- 2¼ lbs thick yogurt or 1 × 1 lb batch of Sweet Labneh, drained weight (page 67)
- pinch of saffron threads, infused in 1 teaspoon warm water
- seeds from 8 cardamom pods (crush the pods and carefully pick out the seeds, discarding the husks)
- ¾ cup superfine sugar
- 2 tablespoons slivered (or nibbed) pistachio nuts

If you're using yogurt rather than labneh, you'll need to strain it first. If you are using labneh, place it in a bowl and skip the next three steps in the recipe.

To strain your yogurt, set a sieve over a large mixing bowl. Drape a large cheesecloth over the surface, so that the base of the sieve is fully lined and the sides of the cheesecloth hang down over the sides of the bowl.

Spoon the yogurt into the center of the cheesecloth, gather up the sides, and tie them tightly with string just above the yogurt. Leave the lengths of string long to tie it up.

Suspend the cheesecloth bag over the mixing bowl by tying it in a position where it can hang and the whey can drip into the bowl. I do this at home by tying the cheesecloth bag onto a cupboard handle over the countertop. Some people like to tie it to a faucet and sit the bowl in the sink below. Set aside in a cool place overnight, during which time the whey will drip through.

The contents of the cheesecloth bag is your labneh. In the morning, put the labneh into a clean earthenware or glass bowl, cover with plastic wrap, and set aside in the fridge until needed.

Pour the saffron infusion over the labneh and stir in with the cardamom seeds and sugar.

Serve chilled, sprinkled with the pistachio nuts.

AMRAKHAND

Similar to the previous recipe for Shrikhand, this version is flavored with mango, a favorite fruit of India. Choose the ripest and most fragrant mangoes you can find for the best flavor. Extra mango can be added for a more intense flavor.

Serves 4

2¼ lbs thick yogurt or 1 × 1lb batch of Sweet Labneh, drained weight (page 67)
pinch of saffron threads, infused in 1 teaspoon warm water
seeds from 8 cardamom pods (crush the pods and carefully pick out the seeds, discarding the husks)
¾ cup superfine sugar
2 medium ripe mangoes, peeled and flesh cut into cubes
2 tablespoons slivered (or nibbed) pistachio nuts

If you're using yogurt rather than labneh, you'll need to strain it first. If you are using labneh, place it in a bowl and skip the next step in the recipe.

To strain your yogurt, set a sieve over a large mixing bowl. Drape a large cheesecloth over the surface, so that the base of the sieve is fully lined and the sides of the cheesecloth hang down over the sides of the bowl.

Spoon the yogurt into the center of the cheesecloth, gather up the sides, and tie them tightly with string just above the yogurt. Leave the lengths of string long to tie it up.

Suspend the cheesecloth bag over the mixing bowl by tying it in a position where it can hang and the whey can drip into the bowl. I do this at home by tying the cheesecloth bag onto a cupboard handle over the countertop. Some people like to tie it to a faucet and sit the bowl in the sink below. Set aside in a cool place overnight, during which time the whey will drip through.

The contents of the cheesecloth bag is your labneh. In the morning, put the labneh into a clean earthenware or glass bowl, cover with plastic wrap, and set aside in the fridge until needed.

Pour the saffron infusion over the labneh and stir in with the cardamom seeds.

Place the sugar and chopped mango in a food processor and process until smooth. (If you don't have a processor, you can mash the mango in a bowl and then strain it through a sieve to form a smooth purée.)

Stir the mango purée into the flavored labneh, cover the bowl with plastic wrap, and chill in the fridge until needed. Sprinkle on the pistachio nuts at the last moment before serving.

BEANS & LEGUMES

HOMEMADE MISO SOUP

Miso is a delicious Japanese paste made from fermented soybeans. After much experimentation, I decided this process is best left to the experts, and I recommend you buy good-quality organic miso, which is sold in supermarkets and online, instead of making it yourself. Always ensure that you buy unpasteurized miso in order to enjoy its full health benefits.

There are three types of dark miso paste to choose from: barley, hatcho, or brown rice miso. Barley miso is made with a blend of barley and soybeans. Hatcho miso is made exclusively from soybeans and is revered for its high protein content. Brown rice miso is made from a blend of brown rice and soybeans.

Miso soup is very straightforward to make and highly nutritious. This version is also delicious with some dried seaweed stirred in at the very end.

Serves 4

4 dried shiitake mushrooms
1½ quarts boiling water
2¼ cups kale, finely chopped
1 large carrot, peeled and very thinly sliced at an angle
¼ cup good quality miso paste (see page 159)

Tip:

Keep the pan off the heat when you add the miso paste. If it boils, it loses much of its goodness.

Start by soaking the mushrooms. Place the mushrooms in a large pan, cover with boiling water, and set aside to soak for about 15 minutes.

Set the pan of mushrooms over medium heat and bring to a boil. Add the kale and carrot and simmer until tender, about 3–5 minutes. Remove the pan from the heat and stir the miso paste into the broth.

Serve immediately.

GRILLED MISO CHICKEN WITH BOK CHOY

This miso marinade adds a wonderful balance of sweet and savory flavors to the chicken. It takes just a few minutes to make and assemble the marinade, and I like to prepare this the night before I want to eat it and leave it covered in the fridge for the next day. The flavor imparted into the chicken is superb, and it makes a very quick and nutritious meal to cook when time is of the essence.

Serves 4

4 heads of bok choy, halved lengthwise
1 tablespoon miso paste
4 scallions, thinly sliced
1 in fresh root ginger, peeled and grated
2 cloves of garlic, peeled and finely chopped
2 tablespoons honey
2 tablespoons soy sauce
4 boneless chicken breasts, skin removed and sliced in half lengthwise
freshly cooked rice, to serve

Bring a saucepan of water to a boil over medium heat and very lightly boil the bok choy for 3 minutes. Drain and refresh in cold water and set aside.

Place the miso, scallions, ginger, and garlic in a bowl. Stir in the honey and soy sauce. Add the chicken and stir to coat it well. Leave to marinate in the fridge for a minimum of 30 minutes, or ideally overnight.

Preheat the broiler or a heat a cast-iron grill pan over high heat. Broil the chicken for about 15 minutes, turning halfway through. Add the bok choy and broil for an additional 5 minutes. Check the chicken is cooked by cutting open a piece in the thickest part and checking there is no pink meat inside.

Serve immediately with freshly cooked rice.

Tip:

My vegetarian recipe testers loved making this with thick (½ in) slices of eggplant instead of chicken.

ROASTED VEGETABLES WITH MISO SAUCE

This is such a lovely vegetable dish. Serve it as a side, perhaps with roast or barbecued meat, grilled tofu, or just as a meal in itself. You can even serve them cold, as a salad. Sticky, caramelized roasted vegetables are surprisingly tasty and prove very popular, even with those who say they don't like vegetables very much.

Serves 4

For the miso sauce

3 teaspoons miso paste
2 teaspoons rice wine vinegar
1 tablespoon soy sauce
3 teaspoons honey

For the vegetables

6 cups prepared vegetables, which could include any mix of eggplant, peppers, carrots, zucchini, parsnips, red onions, and beets, cut into even 1 in chunks (young vegetables can be kept whole)
sea salt and black pepper

To serve

6⅔ cups spinach
1 lime, cut into wedges
Fresh cilantro leaves
2 limes, halved

Preheat the oven to 350°F.

Put the ingredients for the miso sauce in a small bowl and stir together.

Put the prepared vegetables in a large bowl. Pour the miso sauce over the vegetables and toss well to ensure every piece is coated in sauce. Season to taste.

Transfer to a large baking sheet and bake for 25-35 minutes, until the vegetables are tender and caramelized around the edges.

Just before serving, stir the spinach through the vegetables. Serve with lime wedges and some fresh cilantro leaves to garnish.

DOSA PANCAKES WITH DAL, ROASTED CAULIFLOWER AND COCONUT CHUTNEY

Dosa pancakes originate from Southern India, where they often eaten for breakfast. Traditionally the batter is left to ferment for a couple of days before cooking, but this recipe is much quicker as the batter only needs 3–4 hours to ferment at room temperature. The pancakes are delicious eaten with chutney, or as an accompaniment to curry. Some roasted cauliflower and plenty of brinjal (eggplant) pickle are essential accompaniments.

Makes 6 pancakes

For the dosa pancakes

1 cup chickpea flour
1 cup all-purpose flour
½ teaspoon baking soda
1 level tablespoon yellow mustard seeds, crushed gently using a mortar and pestle
1 tablespoon vegetable oil, plus a little extra for cooking
1¾ cups water
pinch of sea salt

For the dal

1 cup red lentils
1 teaspoon sea salt, plus another pinch
1 quart water
3 teaspoons vegetable oil
1 teaspoon finely grated fresh ginger
1½ teaspoons ground turmeric
1 teaspoon ground cumin
1 teaspoon ground coriander
pinch of cayenne pepper

For the roasted cauliflower

1 medium cauliflower, broken into florets
olive oil, for drizzling
sea salt and black pepper

For the coconut chutney

1½ cups fresh coconut flesh (shelled weight), brown skin peeled with a potato peeler
1 very small shallot, chopped
1 teaspoon finely grated fresh ginger
1 tablespoon lemon juice
1 teaspoon ground coriander
2 cloves
½ teaspoon ground cinnamon
1 teaspoon sea salt
1 tablespoon superfine sugar
2 tablespoons chopped fresh cilantro leaves

brinjal pickle, to serve

First, prepare the dosa batter. Put all the ingredients in a measuring cup and whisk to form a smooth batter. Cover with plastic wrap and set aside to ferment for 3–4 hours at room temperature. Preheat the oven to 400°F.

To make the dal, place the lentils and a pinch of salt in a medium saucepan and cover with the water. Bring to a boil over high heat and cook until the lentils are soft, about 10–15 minutes. Drain through a sieve and put into a serving bowl. Heat the oil in a small frying pan over medium heat, add the ginger, salt, and spices and fry for a minute or two until they sizzle and release their aroma; don't let them burn. Drizzle the fried spices over the lentils and stir to combine. (The lentils can be made in advance and kept warm or reheated, as you wish.)

To roast the cauliflower, scatter the florets over a large roasting tray, drizzle over the olive oil and season with salt and pepper. Roast for 20–25 minutes until lightly brown and tender. Serve hot or cold.

To make the coconut chutney, combine all the ingredients in a food processor and process until the coconut is chopped into fairly small pieces. Transfer to a bowl and set aside.

To cook the pancakes, heat a few drops of oil in a large frying pan over high heat and tilt the pan backward and forward to coat the base. Pour in a ladleful of the batter and swirl to just coat the bottom of the pan—the batter is quite thick. Cook the pancake for 2–3 minutes on the first side, until the bottom is golden and lacy, and then flip over with a spatula and cook on the other side for an additional 2–3 minutes. Once the pancake is cooked, place on a baking sheet covered with nonstick parchment paper and transfer to a cool oven to keep warm while you cook the rest. Try not to stack them on the tray.

Serve the pancakes on warmed plates, filled with the dal, roasted cauliflower, and coconut chutney. Accompany with some brinjal pickle.

TEMPEH GADO GADO WITH SATAY TEMPEH

Tempeh is a great source of protein and can be used in a wide range of quick-to-make recipes, making it an excellent standby ingredient to have on hand. While it is not too difficult to make tempeh at home, by cooking soybeans and fermenting them with a Rhizopus starter (available online), it is quite a lengthy process that takes a couple of days. Given that you can buy excellent organic tempeh in supermarkets and online, I'd recommend buying in a good-quality tempeh to cook with at home.

This is my take on a vibrant Indonesian salad. I like to make plenty, so I have enough to enjoy the following day—it's so good!

Serves 4

2 tablespoons sunflower oil
1 × 8 oz pack of tempeh, cut into ½ in cubes

For the satay sauce

½ cup peanut butter (I prefer crunchy, but smooth works too)
1 garlic clove, peeled and crushed
1 teaspoon grated fresh ginger
1 teaspoon ground turmeric
a few drops of Tabasco
1 teaspoon toasted sesame oil
2 tablespoons soy sauce
1 tablespoon honey
2 teaspoons lemon juice
¼ cup coconut milk

For the gado gado

¼ medium white cabbage, thinly sliced
2¼ cups beansprouts
1 cup green beans, blanched in boiling water for 3 minutes, then drained and cooled
½ cucumber, halved lengthwise and thinly sliced on an angle

To serve

a good handful of fresh cilantro leaves
⅓ cup roasted salted peanuts, chopped

Start by making the satay tempeh. Heat the oil in a large frying pan over medium to high heat and fry the tempeh cubes until lightly browned and crisp. Set aside to cool on a couple of sheets of paper towel to absorb any excess oil.

Put all of the ingredients for the satay sauce in a food processor or blender and pulse until smooth. Pour into a bowl, add the cooled tempeh cubes, and gently stir to coat each cube in the sauce. Set aside to marinate for 30 minutes at room temperature.

Meanwhile, put all of the ingredients for the gado gado salad in a large mixing bowl.

Once the tempeh has marinated, pour the cubes of tempeh and sauce over the salad and gently toss well to coat evenly in the satay sauce. Sprinkle with fresh cilantro and peanuts before serving.

TEMPEH STIR-FRY WITH CHILE PEANUT SAUCE

This is possibly the easiest stir-fry sauce ever, made from pantry ingredients, and also the ultimate in fast food, being ready in about 10 minutes flat. It tastes absolutely scrumptious.

Serves 4

1 tablespoon sunflower oil
1 × 8 oz pack of tempeh, cut into ½ in cubes
¼ cup soy sauce
¼ cup crunchy peanut butter
¼ cup sweet chili sauce
2 tablespoons hot water
1 × 9 oz pack of prepared stir-fry vegetables

Heat a large pan or wok over high heat and add the oil. Put in the tempeh cubes and stir-fry until medium brown in color and just starting to crisp around the edges, about 4–5 minutes.

Meanwhile combine the soy sauce, peanut butter, chili sauce and water in a small bowl. Pour the mixture into the pan and add the vegetables, stirring constantly to coat them in the sauce. Continue to stir-fry for an additional 4–5 minutes until the vegetables are heated through. Serve immediately.

CHILE, GARLIC, AND SESAME TEMPEH LETTUCE WRAPS

This is a super quick way to make a delicious light meal using tempeh. The tempeh is cut into cubes and cooked quickly with some punchy yet aromatic flavors. Served in a large lettuce leaf, they make a wonderful appetizer, starter, or snack. Serve with some extra sweet chilli sauce on the side, either for dipping, or for drizzling on top of the tempeh.

Serves 4

- juice of 1 lemon
- 3 tablespoons soy sauce
- 1 teaspoon toasted sesame oil
- 1½ tablespoons honey
- 2 teaspoons sweet chili sauce
- 1 tablespoon finely grated fresh ginger
- 2 garlic cloves, peeled and crushed
- 1 × 8 oz pack of tempeh, cut into ½ in cubes
- 2 teaspoons sunflower oil
- 4 large, crisp lettuce leaves, such as Iceberg
- 2 carrots, cut into very thin matchsticks
- 3 teaspoons toasted sesame seeds

Whisk together the lemon juice, soy sauce, sesame oil, honey, sweet chili sauce, ginger, and garlic in a large mixing bowl. Add the cubes of tempeh and stir to coat them in the sauce. (If you wish you can prepare the recipe to this stage in advance and leave the tempeh to marinate for anything from 1 hour to overnight, but this isn't essential.)

Heat a large frying pan over high heat and add the oil. Put in the marinated tempeh cubes and stir-fry until medium brown and just starting to crisp around the edges, about 4–5 minutes.

To serve, divide the tempeh between the lettuce leaves and top with the carrot slices and a sprinkling of sesame seeds. Wrap the lettuce around the filling and enjoy immediately.

FERMENTED BLACK BEAN SAUCE

Fermented black beans, or *douchi*, are an important ingredient in Chinese cooking. Made from black soybeans, fermented using salt and then dried, they are completely different to the black beans widely used in Mexican cuisine. This is another occasion when it is a lot easier to buy in fermented black beans rather than fermenting them yourself, and the results will be very reliable, too.

Fermented black beans (also sold as Chinese black beans or salted black beans) are widely sold in Asian supermarkets and are extremely inexpensive, especially as a little goes a long way. They are available in both packages and jars, and I'd recommend you to select the package version, because they are much more flavorful. Due to their intense salty flavor, fermented beans should be generally rinsed before cooking.

This delicious recipe for homemade black bean sauce can be used in all manner of Chinese stir-fries instead of store-bought sauce. It will serve four people, and is particularly good served with 1 lb of chicken, pork, or firm tofu, cut into 1 in cubes.

Serves 4

2 tablespoons fermented black beans
3 garlic cloves, peeled and finely chopped
1 in piece of fresh ginger, peeled and finely grated
1 fresh red chile, finely chopped
4 scallions, thinly sliced
3 tablespoons Shaoxing rice wine
6 tablespoons soy sauce
2 tablespoons toasted sesame oil

you will need a ½-quart glass Le-Parfait-style jar with a lid, sterilized according to the instructions on page 11

Start by soaking the black beans. Put in a small dish, cover with hot water and leave to soak for 5 minutes. Drain and set the beans aside. Add the remaining ingredients and stir together well.

The sauce is ready to heat. It only needs to be poured into a pan and heated for 3–4 minutes to cook it through. It can be served hot and poured onto meat, tofu, or rice, or simply stir-fry strips of chicken, pork or cubes of firm tofu until cooked, and pour in the sauce. Continue to cook on medium heat for an additional 3–4 minutes until hot. Serve immediately.

The black bean sauce can be kept in a jar and stored in the fridge. Use within a week.

MUSHROOMS AND TOFU IN GARLIC BLACK BEAN SAUCE

This is a gorgeous Chinese dish with a wonderfully rich, deep, savory flavor. It can be prepared in 20 minutes, from start to finish and is excellent served with rice, noodles, roasted chicken, or bok choy, or even on its own. You can use any combination of mushrooms you like; cremini, shiitake, and oyster are all delicious.

Serves 2, generously

1 tablespoon toasted sesame oil
3 cups mushrooms, sliced approx. 1 in thick
3½ oz firm tofu, cut into ½ in cubes
1 tablespoon soy sauce
¾ cup hot chicken or vegetable stock

For the sauce

1 tablespoon fermented black beans, rinsed
1 tablespoon soy sauce
2 cloves garlic, peeled and crushed
1 red chile, seeded and finely chopped
1 tablespoon Shaoxing rice wine

Start by making the sauce. Combine all the ingredients in a bowl and process, either using an immersion blender, or place in a food processor. Ensure that the sauce is only lightly blitzed to combine the ingredients, but retains some texture. It should not be processed into a smooth sauce. Set aside.

Heat a large pan or wok over high heat. Add the oil, and when the oil is hot, add the mushrooms and tofu and stir-fry for 2–3 minutes until the mushrooms start to soften. Add the black bean sauce, soy sauce, and stock and cook over high heat, stirring frequently for around 5 minutes until the sauce thickens and reduces. Serve immediately.

Tip:

Rinse the fermented black beans in a sieve in cold running water. Shake off any excess water and set aside.

STEAK WITH FERMENTED BLACK BEANS

This is quite different to anything you'll get from a Chinese restaurant; richly flavored, juicy steak, a deeply flavored, light sauce, and it's really good for you too.

Serves 4

- 2 tablespoons fermented black beans (you can add more to taste)
- 1½ in piece of fresh ginger, peeled and finely grated
- 4 cloves of garlic, peeled and finely chopped
- 2 scallions, thinly sliced at an angle
- 1 fresh red chile, seeded and finely chopped
- 2 tablespoons Shaoxing rice wine or dry sherry
- 1 tablespoon soy sauce
- 6 teaspoons toasted sesame oil
- 1 teaspoons honey
- 1 lb sirloin steak, thinly sliced
- A small head of broccoli, cut into small florets
- 2 tablespoons sunflower oil

Tip:

The steak can be marinated overnight for a fuller flavor. Marinating the meat in advance also makes this a super-quick supper to assemble.

Put the fermented black beans into a bowl and cover with boiling water. Allow to soak for 5 minutes. Drain. Put the beans into a bowl with the ginger, garlic, scallions, chile, rice wine or sherry, soy sauce, sesame oil, and honey. Add the sliced steak, and leave to marinate for at least an hour, or overnight, in the fridge.

Put the broccoli into a pan and cover with boiling water. Blanch for 2-3 minutes, and then drain well.

Heat the sunflower oil in a wok over high heat. Add the steak and stir-fry for 2-3 minutes; depending on how well cooked you like your beef.

Add the broccoli and marinade. Stir-fry for another minute or two and serve immediately.

STIR-FRIED CHICKEN WITH NOODLES AND BLACK BEAN SAUCE

This is a scrumptious stir-fry that doesn't require too much chopping and preparation—indeed, it can be made in under 15 minutes without too much difficulty. Perfect for those days when you just want a big bowl of steaming hot noodles.

SERVES 4

9 oz fine egg noodles
1 tablespoon sunflower oil
2 tablespoons sesame oil
2 large chicken breasts, sliced thinly
4 garlic cloves, chopped
1 in piece fresh ginger, peeled and grated
2 oz fermented black beans
14 oz bok choy, sliced thickly
1 red chile, sliced
6-8 tablespoons soy sauce, to taste
4 scallions, sliced thinly at an angle
2 tablespoons toasted sesame seeds

Cook the noodles according to the package instructions. Drain once cooked.

Heat the sunflower oil and 1 tablespoon of sesame oil in a wok over high heat, then add the chicken and stir-fry for about 5 minutes.

Add the garlic, ginger, and black beans and stir-fry until fragrant, followed by the bok choy and chile.

Add the noodles to the wok, then the soy sauce. Toss until piping hot.

Serve in bowls, with the scallions and sesame seeds sprinkled on top. You may wish to add more soy sauce to taste.

TIP:

This stir-fry keeps well. Any leftovers will make a fine lunch the following day.

SOURDOUGH BAKING

SOURDOUGH BREAD

Sourdough bread is one of the most rewarding things to bake. It takes a bit of practice to get it right, but the results are delicious. Don't be put off if it doesn't work perfectly first time; make a note of what you did and eventually you'll perfect your loaf, getting better every time you make it.

Sourdough bread is simply made from flour, water, and salt—with no added yeast, unlike ordinary bread. Instead, it slowly harnesses wild yeasts in the atmosphere to activate these simple ingredients. This means conditions such as the environment you live in, and the temperature of your home, all play a part in the process—which is why you may need to tweak the method to work for you.

Sourdough bread not only tastes fantastic, but it is also easier to digest than most other breads. This is because the process of fermentation digests the phytic acid in the grain. Phytic acid, or phytate, is a substance found in many plant foods, including grains. When ingested, it binds nutrients in the digestive tract, which inhibits their absorption by the body. Fermenting grains, such as wheat flour, by making a sourdough starter breaks down the phytic acid, making vitamins and minerals in the grain more available for our bodies to absorb.

Always use organic flour for best results. If you're on muncipal water, it's a good idea to use filtered water, if you can, because chlorine and chemicals could inhibit the fermentation process.

To make sourdough bread, you will first need to make a starter. Sometimes, people are given a starter from a friend, who already has one established. If you're not so lucky, and don't want to make your own starter, you could buy the starter from a specialty bakery or delicatessen (see page 159 for resources). If this is the case, you can skip to Step 2 of the recipe.

WHITE SOURDOUGH BREAD

If you've never made sourdough before, this is the place to start. It can take a couple of attempts at sourdough to get it right, but that's the same for everyone, so don't be disheartened if your first attempt doesn't turn out perfectly. Sourdough is a living, breathing thing, and it reacts differently, depending on the weather, where in the world you live, and the ingredients you use. You can either make your starter from scratch, or use some from someone who has one already established, which saves a couple of days. If you're given some active starter, you can skip to Step 2.

MAKES APPROX. 14 OZ OF STARTER

- 2 cups organic white bread flour (added over 4 days)
- ⅔ cup filtered or bottled mineral water at room temperature (see page 8)

STEP 1—MAKING THE SOURDOUGH STARTER

Day 1
Measure out ¾ cup of the flour and ½ cup of the water in a large mixing bowl. Stir well with a wooden spoon to form a smooth paste without any lumps. Loosely cover the bowl with plastic wrap or a cloth and set aside on your kitchen countertop for 3 days. Avoid disturbing it during this time.

Day 4
After 3 days, the starter will be runnier and smell slightly fruity or vinegary. Stir the starter with a wooden spoon and mix in 2 tablespoons of water, stirring well to evenly incorporate, then ⅔ cup flour. Cover the bowl lightly with plastic wrap or a cloth and set aside for another day on the countertop.

Day 5
Weigh the mixture and throw away half. Place the remaining 7 oz (approx.) back in the bowl and stir in the remaining ⅔ cup flour and 2 tablespoons water. Cover with plastic wrap or a cloth and set aside on the kitchen countertop for an additional day.

Day 6
Your sourdough starter is now active and ready to use. It should smell fruity and vinegary, but not excessively sour. If the starter smells very strong, and possibly unpleasant, it's likely it has died, and you will need to start again. There are many reasons for starters dying, and it can sometimes be caused by ambient temperature or the water you're using. Later on, it can die by being fed too little. The best advice I can give you is not to worry. It happens. The hardest part of making a starter is getting it going. Just throw it away and start again. Keep a note of everything you do, because it may help identify future successes and failures further down the line.

Weigh out 3 oz of the starter and store in a sealed plastic container in the fridge. This is your sourdough insurance policy. You can leave it indefinitely in the fridge or freezer as a back-up. Don't worry if it separates. The vinegary layer that forms on the surface is called hooch. It's a protective layer, and it may turn black in color, but this is absolutely fine. When you are ready to use it, simply stir the hooch into the solids and use in Step 2 of the recipe.

Makes approx. 1 lb levain

For the levain

3 oz sourdough starter (see page 106)
¾ cup filtered or mineral water at room temperature (see page 8)
1⅔ cups organic white bread flour

Makes 1 large loaf (approx. 2¼ lbs)

5¼ cups organic white bread flour
¾ cup dark rye flour
2 cups lukewarm water
6 oz Sourdough Levain (see above)
1 tablespoon (15g) fine sea salt

you will need a well-floured 2¼ lb banneton (see page 11)
a sharp knife, grignette or lame to slash the loaf

STEP 2—MAKING THE LEVAIN

A levain is the stage between the starter and the bread dough. You take a small quantity of the sourdough starter and activate it. Sourdough levain is what I use in all of my sourdough baking recipes.

Place 3 oz sourdough starter in a large mixing bowl and stir in the water. Add the flour and mix to a thick paste. Cover the bowl loosely with plastic wrap and set aside at room temperature for at least 4 hours, after which time it will start to bubble. This shows it's ready to bake with. The levain will remain active for up to 3 days, stored covered in plastic wrap on the kitchen countertop; just make sure you stir it well before use. After 3 days, it will become thin and runny again. At this point, keep back 3 oz to use as a starter for your next batch of levain. (As before, you will need to mix it with ¾ cup water and 1⅔ cups flour and then set it aside on the kitchen countertop for between 4 hours and 3 days until needed.)

STEP 3—MAKING THE SOURDOUGH BREAD

Put all the ingredients in a large mixing bowl and mix together to form a wet and sticky dough using a spatula or wooden spoon. Set aside at room temperature for about 20 minutes, covered lightly with plastic wrap. Knead gently in the bowl for 2–3 minutes and then set aside for 1 hour, covered with plastic wrap.

By this time, the dough will still be quite wet and sticky. It won't increase in size yet. Don't worry if you can still see streaks of flour through the dough. You want to treat your sourdough gently, rather than knead it, as you would with yeasted bread, as kneading will produce a tough loaf. Turn the dough out onto a floured work surface. Lift one edge of the dough and fold it into the center. Give the dough a quarter turn and repeat this process 4 times. This will start to bring your dough into a ball.

Place the dough, rounded side down, seam side up, into a very well-floured banneton and cover with a damp dish towel (this helps to prevent the bread from drying out). Transfer the bread to a sunny spot or warm cupboard and set aside to ferment for 4–6 hours, during which time the dough should rise to the top of the banneton. (You can leave the bread to ferment for up to 12 hours, which will result in a slightly more sour flavor.

Preheat the oven to 400°F and place a baking sheet or baker's stone inside the oven to heat up. Just before you are about to bake your loaf, remove the baking sheet or stone from the oven and dust it liberally with flour. Turn the bread onto it and deeply score the surface with a sharp blade. Bake for 40–50 minutes until the crust is brown and firm, and the loaf feels lighter, crisp, and firm. Remove from the oven and transfer to a wire rack to cool completely.

This bread keeps very well, and is delicious toasted when a few days old.

RYE SOURDOUGH

This is my take on Scandinavian-style rye bread. I have had the best results by adjusting the quantities of flour used in the white sourdough loaf (see opposite), using the same levain for both sourdough bread recipes in this book. I like to use a traditional stoneground rye flour here. Many small, independent flour mills produce their own varieties, which I like to use, as they are often organic, unbleached, and milled using fewer industrial processes.

Makes 1 medium loaf (approx. 2¼ lbs)

2½ cups organic white bread flour
3½ cups dark rye flour
2 cups lukewarm water
6 oz Sourdough Levain (page 108)
1 tablespoon (15g) fine sea salt

you will need a well-floured 2¼ lb banneton (see page 11)
a sharp knife, grignette or lame to slash the loaf

Put all the ingredients in a large mixing bowl and mix together to form a wet and sticky dough using a spatula or wooden spoon. Set aside at room temperature for about 20 minutes, covered lightly with plastic wrap. Knead gently in the bowl for 2–3 minutes and then set aside for 1 hour, covered with plastic wrap.

By this time, the dough will still be quite wet and sticky. It won't increase in size yet. Don't worry if you can still see streaks of flour through the dough. You want to treat your sourdough gently, rather than knead it, as you would with yeasted bread, as kneading will produce a tough loaf. Turn the dough out onto a floured work surface. Lift one edge of the dough and fold it into the center. Give the dough a quarter turn and repeat this process 4 times. This will start to bring your dough into a ball.

Place the dough, rounded side down, seam side up, into a very well-floured banneton and cover with a damp dish towel (this helps to prevent the bread from drying out). Transfer the bread to a sunny spot or warm cupboard and set aside to ferment for 4–6 hours, during which time the dough should rise to the top of the banneton. (You can leave the bread to ferment for up to 12 hours, which will result in a slightly more sour flavor.

Preheat the oven to 400°F and place a baking sheet or baker's stone inside the oven to heat up. Just before you are about to bake your loaf, remove the baking sheet or stone from the oven and dust it liberally with flour. Turn the bread onto it and deeply score the surface with a sharp blade. Bake for 40–50 minutes until the crust is brown and firm, and the loaf feels lighter. Remove from the oven and transfer to a wire rack to cool completely.

SOURDOUGH SAJ

These Middle Eastern flatbreads are widely enjoyed across the region. On my first trip to the Middle East, they were one of my favorite foods, and I always make a point of seeking them out when I return.

There's nothing quite like a hot, freshly cooked flatbread, so I urge you to go to the effort of making them. I really like them sprinkled with za'atar and cut into triangles. Served hot, they are extremely popular as an appetizer.

MAKES 6–8, DEPENDING ON THICKNESS AND SIZE

5 cups all-purpose flour, sifted
½ teaspoon sea salt
5 teaspoons superfine sugar
1½ cups lukewarm water
2 oz Sourdough Levain (page 108)
1 tablespoon olive oil
Za'atar (page 78)

Put the flour, salt, and sugar in a large mixing bowl and mix together briefly.

Add the water, sourdough levain, and olive oil and mix well to form a stiff dough. The dough may seem a bit dry, but don't be tempted to add more water—just keep going until the dry ingredients are incorporated. Knead the dough for 10 minutes, until smooth. This will be quite hard work, so I recommend using a stand mixer fitted with a dough hook if you have one.

Once the dough has been kneaded, transfer it to a large mixing bowl and cover with plastic wrap. Then set aside in a warm place to rest for at least 30 minutes. (You can make it in advance and leave it covered in plastic wrap for up to 12 hours, if you like). It won't rise as dramatically as other breads, but it should puff up a little.

Heat a large, nonstick frying pan or grill pan over high heat. Pinch off a little ball of dough, about the size of a plum, and roll out on a floured work surface. Roll as thinly or thickly as you like, but ensure it is no thicker than ⅓ in. I prefer them rolled very thin.

Place the flatbread in the hot pan and cook quickly on both sides until lightly spotted, about 2 minutes on each side. Keep the saj warm by wrapping them in a dish towel while you cook the rest.

When the saj is nearly cooked, brush one side with extra virgin olive oil, sprinkle with homemade za'atar (see page 78), and flip over to cook for 10 seconds before removing from the pan. Serve immediately.

SOURDOUGH HOT CROSS BUNS

Sourdough can be used in all manner of cakes and traditional yeasted doughs. I love to put some sourdough in hot cross buns, which gives them a little extra depth of flavor.

MAKES 12–16

FOR THE BUNS

3¾ cups white bread flour, plus extra for dusting
1 teaspoon sea salt
½ cup superfine sugar
1 level teaspoon ground cinnamon
1 level teaspoon mixed spice
2 teaspoons instant action dried yeast
finely grated zest of 1 orange
finely grated zest of 1 lemon
⅔ cup golden raisins
1¼ cups whole milk, heated until lukewarm
6 tablespoons salted butter, melted
1 large egg, beaten
2 oz Sourdough Levain (page 108)

FOR THE CROSSES

5 tablespoons Sourdough Levain (page 108)
5 tablespoons all-purpose flour
3 teaspoons water

FOR THE GLAZE

1 egg, beaten

Put the flour, salt, sugar, spices, yeast, orange and lemon zest, and sultanas in a large mixing bowl and stir to combine.

Add the warm milk, melted butter, egg, and sourdough levain and mix well using a wooden spoon or spatula to form a sticky dough.

Either place the dough onto a lightly floured work surface and knead by hand for 5 minutes, or use a stand mixer fitted with a dough hook instead. Place the dough back in the bowl, cover with plastic wrap, and set aside to rise in a warm place for 1½–2 hours or until doubled in size.

Place the dough onto a lightly floured work surface and cut into 12–16 equal-sized pieces, depending on whether you want small or large buns. Shape into rounds. Arrange the buns butting up to one another on a large floured baking sheet. Cover with oiled plastic wrap and set aside to prove in a warm place for about an hour or until well risen.

Meanwhile, mix together all of the ingredients for the cross paste in a small bowl and set aside. Preheat the oven to 400°F. Once the buns have risen nicely, you can pipe on the crosses using a piping bag fitted with a small-sized round nozzle, or a homemade paper-piping bag. Carefully brush the tops with beaten egg. Bake for 15 minutes, until they are lightly browned, glossy, and feel light. Remove the buns from the oven and set aside to cool on a wire rack. Enjoy split and buttered.

These are delicious toasted the next day.

SOURDOUGH CHRISTMAS CAKE

This is a delicious simple Christmas cake. It's really easy to make, and a good place to start if you've never made Christmas cake before.

Serves 16–20

For overnight soaking

6 cups dried fruit; a mixture of golden raisins, raisins, and currants, or just golden raisins
3½ oz candied orange or lemon peel (or a mixture of both), chopped
9 oz Glacé (candied) cherries
5½ oz brandy, whiskey, or dark rum

For the cake

1 cup salted butter, softened
1½ cups dark brown muscovado sugar
4 large eggs, beaten
pinch sea salt
1 teaspoon ground ginger
½ teaspoon ground allspice
½ teaspoon ground cinnamon
½ teaspoon ground coriander
zest of 1 large orange
3 tablespoons ground almonds
1¾ cups self-rising flour, sifted
2 teaspoons baking powder
3 oz Sourdough Levain (page 108)

you will need a 8 in spring-form cake pan, approx. 4 in deep

Start by soaking the fruit. Put the dried fruit, candied peel, glacé cherries, and brandy in a large mixing bowl and stir well to combine. Cover the bowl with plastic wrap and set aside on the kitchen countertop overnight. (You can get away with about 4 hours soaking, but overnight is best.)

The next day, line your cake pan with parchment paper and preheat the oven to 350°F.

Put the butter and sugar in a very large mixing bowl and beat together until light and fluffy. Beat in the eggs a little at a time, beating well after each addition.

Add the salt, spices, orange zest, and almonds and stir well, followed by the soaked fruit and all of the soaking liquid and beat well together. Finally, add the sifted flour, baking powder, and sourdough levain. Fold in gently until evenly combined.

Spoon the mixture into the prepared pan and neatly level the surface. Bake for 2–2½ hours, or until a skewer inserted into the center of the cake comes out clean. Check your cake after 2 hours and see if it's done. If your oven is a little slow, the cake may need an additional 15–30 minutes, in which case cover it loosely with foil to stop the top from browning too much.

Remove the cake from the oven and set aside to cool in the pan for 20–30 minutes before turning it onto a wire rack. Remove the parchment paper and set aside to cool completely before wrapping and storing or decorating.

This cake can be eaten the same day (unlike other Christmas cakes that need maturing or feeding). It also keeps very well, stored in an airtight container for up to a month before cutting, and once cut, eat within 2 weeks, ensuring it's well-wrapped when stored.

SOURDOUGH STOLLEN

This is my version of a traditional stollen. The dough is light, sweet, and airy, and packed with delicious fruit and cubes of marzipan. This is a good bake to make to share with a crowd, because stollen doesn't keep for long once made.

Makes 1 large loaf (approx. 2 lbs)

For the stollen

4 cups all-purpose flour, plus a little extra for the work surface
Pinch salt
1 cup superfine sugar
1 tablespoon instant action dried yeast
1 cup milk
½ cup salted butter
1 egg, beaten
3 oz Sourdough Levain (page 108)
1 cup golden raisins
1 heaped teaspoon cinnamon
finely grated zest of 1 lemon
finely grated zest of 1 orange

For the filling

4½ oz marzipan, cut into ½ in cubes
6 oz candied orange or lemon peel (or a mixture of both)
¾ cup glacé (candied) cherries

For the topping

1 egg, beaten
2 tablespoons confectioner's sugar, sifted

Start by making the stollen dough. Sift the flour into a large mixing bowl. Add the salt, sugar, and yeast and stir well.

Put the milk and butter into a saucepan over low heat and melt the butter into the milk. Do not allow the milk to start to simmer; this needs to be done over a low temperature.

Pour the milk and melted butter into the flour mixture, add the egg, sourdough levain, golden raisins, cinnamon, lemon, and orange zest and continue to stir together to form a soft dough.

Grease a large mixing bowl; place the dough into the bowl. Cover with plastic wrap and put in a warm place for at least 4 hours until the dough has increased in size by about a third in size. You can even leave this dough overnight.

When the dough has risen, remove from the bowl, and place on a clean, floured work surface. Gently roll out the dough to form a rectangle around 1 in thick, no thinner.

Evenly sprinkle the marzipan, candied peel, and cherries over the dough. Gently roll up, rolling from one long side to another, so that you have a long sausage.

Place, seam downward, on a floured baking sheet. Tuck in the ends underneath. Cover and transfer to a warm place to prove for another hour. The dough will not rise dramatically.

Preheat the oven to 325°F. Brush the stollen with the beaten egg and bake for 45 minutes to 1 hour, until browned and firm.

Remove from the oven and cool on a wire rack. When cool, drench in sifted confectioner's sugar. Slice and serve. Enjoy within 2 days.

SPICED APPLE SOURDOUGH CRUMBLE CAKE

This is a superb cake for every day, filled with juicy chunks of apple, flavored with a hint of spice, and finished with a sweet crumble topping. If you wish you can bake the mixture in individual muffin cases. There should be enough mixture for 12–16 muffins, depending on the size of your cases.

Serves 10

For the topping

⅓ cup superfine sugar
⅔ cup all-purpose flour
5 tablespoons cold salted butter, cut into cubes

For the cake

2¼ cups self-rising flour
1 teaspoon baking powder
¾ cup superfine sugar
pinch of fine sea salt
1 teaspoon mixed spice
½ cup salted butter, cut into cubes, plus extra for greasing
1 cup whole milk
1 teaspoon vanilla extract
2 large eggs, beaten
3½ oz Sourdough Levain (page 108)
1¾ cups apples, peeled, cored, and cut into ½ in cubes (prepared weight)

you will need a 8 in round, spring-form cake pan, approx. 3 in deep

Preheat the oven to 350°F. Grease your cake pan, line the base and sides with non-stick parchment paper, and set aside.

To make the crumble topping, put the sugar and flour in a medium bowl and stir together. Add the butter and rub in with your fingertips until the mixture resembles sticky breadcrumbs. If your butter is soft it will form clumps, but that's fine. Alternatively, process the ingredients together in a food processor to form rough breadcrumbs. Don't over-mix, or the crumble mixture will come together to form a ball.

To make the cake, sift the flour and baking powder into a large mixing bowl. Add the superfine sugar, salt, and mixed spice and stir to combine. Add the cubes of butter and rub in with your fingertips until the mixture resembles breadcrumbs.

In a separate bowl, whisk together the milk, vanilla extract, and eggs. Pour the wet ingredients into the dry ingredients and add the sourdough levain. Gently stir the ingredients together to form a smooth batter, but don't over-mix or the cake will be tough. Finally, add the apple chunks and stir in gently to combine.

Spoon the mixture into the prepared pan and top with the crumble topping. (If your crumble topping has formed large clumps, just break them up with a knife and place small lumps of topping on top of the cake batter.)

Bake for 40-50 minutes, until well-risen and golden brown. (If using muffin cases, allow 22-28 minutes, depending on size.) Transfer the cake to a wire rack and set aside to cool.

This cake is best enjoyed as fresh as possible.

SOURDOUGH CARROT CAKE

This is a really fruity carrot cake, packed with lots of lovely, juicy ingredients. I've left out the nuts, but if you can't live without them in your carrot cake, just add 3 oz chopped pecans or walnuts. This recipe makes one large cake to feed a gathering.

Serves 14

3 cups coarsely grated carrots (peeled before grating)
2 oz dried coconut
1 × 7 oz can of pineapple, drained and chopped
⅔ cup golden raisins
2 cups brown sugar
¾ cup softened salted butter, plus extra for greasing
3 large eggs, beaten
1 teaspoon vanilla extract
3¼ cups all-purpose flour
1 teaspoon baking soda
2 teaspoons ground cinnamon
1 teaspoon mixed spice
½ teaspoon fine sea salt
5½ oz Sourdough Levain (page 108)

For the Cream Cheese Frosting

¼ cup unsalted butter, softened
generous ½ cup full fat cream cheese
2 cups icing sugar
scant ¼ cup dried cranberries
¼ cup pecan nuts
finely grated zest of 1 lime
scant ¼ cup poppy seeds

you will need an 11 in round, spring-form cake pan, approx. 3 in deep

Preheat the oven to 350°F and grease and line your 11 in round cake pan with nonstick parchment paper. Combine the carrot, coconut, pineapple, and golden raisins in a mixing bowl, stir well and set aside.

In a separate, large mixing bowl or stand mixer, beat together the sugar and butter until light and fluffy. Add the beaten eggs a little at a time, beating well after each addition. Stir in the vanilla extract. Sift the flour, baking soda, cinnamon, and mixed spice into the cake mixture and then add the salt. Carefully fold everything together to form a thick cake batter.

Add the sourdough levain and stir in carefully, followed by the carrot mixture. Give the mixture a final stir to ensure all the ingredients are combined, and then spoon into the prepared pan. Bake for 40–50 minutes, until well risen and golden brown. To check it is cooked, insert a toothpick in the center of the cake—it should come out clean. Remove the cake from the oven and set aside to cool in the pan for 10 minutes before transferring to a wire rack to cool completely.

To make the frosting, put the butter and cream cheese into a bowl and beat together. Sieve the icing sugar into a separate bowl, and add to the butter and cream cheese a few spoons at a time. Beat the sugar in well. When all the sugar has been incorporated, continue to beat the mixture for a further 2 minutes to ensure it is smooth before frosting your cake.

Ice the cake with the cream cheese frosting and top with the dried cranberries, pecan nuts, poppy seeds, and lime zest.

Store in an airtight container and consume within a week.

SOURDOUGH CHOCOLATE CAKE

This is a particularly good chocolate cake. It's very chocolatey, moist and rich, and makes a wonderful celebration cake. Or, just make it for no reason. It keeps well.

Serves 8–10

For the cake

1½ cup brown sugar
1 cup softened salted butter
pinch of fine sea salt
3 large eggs, beaten
1 teaspoon vanilla extract
2 oz Sourdough Levain (page 108)
1 cup cocoa powder
2 cups all-purpose flour
2 teaspoons baking powder
1 cup whole milk
2 oz chocolate of your choice (see Tip, below), finely chopped

For the icing

⅔ cup salted butter, softened
½ cup cocoa powder
1⅔ cups confectioner's sugar
1–2 tablespoons whole milk

you will need 2 × 8 in round cake pans, at least 2 in deep

Preheat the oven to 400°F and grease and line your cake pans with non-stick parchment paper.

Combine the sugar and butter in a large mixing bowl and beat together really well, until pale and fluffy. Add the salt, eggs, and vanilla extract, and beat again. Stir in the sourdough levain. Sift in the cocoa, all-purpose flour, and baking powder, and fold in carefully. Finally, stir in the milk and chocolate chunks.

Divide the mixture between the prepared pans and smooth the surface. Bake for 25–35 minutes, or until the cakes are well risen and a toothpick inserted into the center comes out clean.

Remove the cakes from the oven and set aside to cool in their pans for 10 minutes before turning them onto a wire rack to cool completely.

To make the icing, place the butter in a mixing bowl and sift over the cocoa and confectioner's sugar. Beat the ingredients together with a wooden spoon or electric beaters, adding the milk a little at a time to form a smooth, spreadable icing.

Sandwich the cakes together with half the icing and spread the rest over the top.

Store in an airtight container and eat within a week.

Tip:

Rather than buying chocolate chips, which tend to be made from very poor-quality chocolate, I buy good-quality dark or milk chocolate and chop it into little pieces by hand.

SOURDOUGH CHOCOLATE MUFFINS

These dark, intensely chocolatey muffins are very popular with grown-ups. The sourdough makes them slightly less sweet than commercial muffins, which, for me, is a good thing. I always buy bars of chocolate and chop them up with a sharp knife, rather than buying chocolate chips, because the quality and flavor is so much better. You can use dark, milk, or white chocolate here, whichever you prefer.

MAKES 12–16, DEPENDING ON THE SIZE OF YOUR CASES

2¼ cups self-rising flour
1 teaspoon baking powder
¾ cup cocoa powder
1½ cups brown sugar
½ teaspoon fine sea salt
1 cup whole milk
½ cup salted butter, melted and cooled slightly
2 large eggs, beaten
1 teaspoon vanilla extract
3½ oz Sourdough Levain (page 108)
2½ oz chocolate of your choice, broken into chunks
approx. ¼ cup demerara sugar, for sprinkling

you will need 1–2 muffin pans, along with 12–16 muffin cases

Preheat the oven to 350°F and line your muffin pan/s with paper cases.

Sift the flour, baking powder, and cocoa into a large mixing bowl. Add the sugar and salt, and stir together until evenly combined.

Measure out the milk into a measuring cup and whisk in the melted butter, eggs, and vanilla extract. Pour the wet ingredients over the dry ingredients in the bowl and add the sourdough levain. Gently stir the ingredients together to form a smooth batter. You want to ensure that all of the dry ingredients are incorporated into the batter, but don't over-mix or the muffins will be tough. Finally, add the chocolate chunks and fold in carefully.

Spoon the mixture into the prepared muffin cases and top with a sprinkling of demerara sugar, allowing approx. ½ teaspoon per muffin.

Bake for 20–25 minutes, depending on size, or until the muffins are well risen and a deep golden brown. Set aside to cool on a wire rack.

Store in an airtight container and eat within 3 days. These muffins freeze very well.

SOURDOUGH CINNAMON BUNS

These buns are my take on the delicious sweet buns I have enjoyed in Scandinavia. They make a very special breakfast—I like to make them on Christmas Day and birthdays—or they are delicious served with coffee.

Makes 9 buns

For the buns

1¾ cups all-purpose flour, plus a little extra for dusting
1⅔ cups white bread flour
pinch of sea salt
⅓ cup superfine sugar
2 level teaspoons instant action dried yeast
1 cup whole milk
5 tablespoons salted butter
1 large egg, beaten
2 oz Sourdough Levain (page 108)

For the filling

6 tablespoons softened salted butter
⅓ cup demerara sugar
4 teaspoons ground cinnamon

For the topping

1 large egg, beaten
3 tablespoons demerara sugar

you will need a 8 in square or round cake pan, approx. 3 in deep

Sift the flours into a large mixing bowl. Add the salt, superfine sugar, and yeast and stir well.

Combine the milk and butter in a small saucepan and set over low heat until the butter has melted. Do not allow the mixture to simmer; this needs to be done over a low temperature.

Pour the milk and melted butter mixture over the dry ingredients, add the egg and sourdough levain, and stir together to form a soft dough. Grease a large mixing bowl and place the dough inside. Cover the bowl with plastic wrap and set aside to rise in a warm place for about 45 minutes, or until the dough has doubled in size.

Meanwhile, make the filling by beating together the butter, demerara sugar, and cinnamon. Grease your cake pan and line with non-stick parchment paper. Set aside.

Place the dough onto a floured work surface and gently roll it out to form an 8 × 18 in rectangle, approx. 1 in thick (no thinner). Spread the filling evenly over the surface of the dough.

Starting at the longest edge, roll up the dough to form a Swiss-roll shape and cut into 9 slices, approx. 2 in thick. Arrange the buns inside the prepared pan and cover the surface with oiled plastic wrap. Set aside to prove in a warm place for an additional 30 minutes.

Preheat the oven to 350°F.

Brush the tops of the buns with beaten egg, sprinkle over some demerara sugar, and bake for 25–35 minutes, until golden brown.

Remove the buns from the pan, without separating them, and set aside to cool on a wire rack for 30 minutes before pulling them apart and diving in.

These buns are best eaten on the day they are made.

DRINKS

WATER KEFIR

Water kefir is a wonderfully light and refreshing fermented drink, full of natural probiotics. This is a great drink to start with if you're not used to fermented drinks because it's very mildly flavored.

MAKES 1 QUART

1 quart filtered water (see page 8)
2 tablespoons water kefir grains (available online)
3 tablespoons organic granulated cane sugar

you will need a 1-quart glass Le-Parfait-style jar with a rubber seal, sterilized according to the instructions on page 11

Put all of the ingredients in the jar and stir well with a wooden (not metal) spoon. Set aside to ferment on your kitchen countertop for 2–3 days. You can either leave the lid on or off as the kefir ferments. If you leave the lid on, you will notice the kefir becomes slightly fizzy.

You don't need to do much to the kefir as it ferments, apart from giving it the odd shake, to allow the sugar to disperse more easily.

Strain through a nylon sieve into a clean pitcher, reserving the kefir grains to use again (see next recipe). I recommend starting with a small glass of kefir at first, roughly ½ cup. It's nicest served chilled.

TIPS:

The grains feed on the sugar in the kefir and multiply as the drink ferments, so don't be alarmed if you end up with more grains than you started with—that's a good sign!

You can reuse the water kefir grains a second time to make an extra batch. There is no need to rinse them; use them as they are. They can be stored in the fridge in a clean glass or plastic container.

FRUIT WATER KEFIR

After your first attempt, you might like to experiment by adding other flavors to your water kefir on page 126. These are some of my favorite flavor combinations, which add a subtle fruity flavor to your drinks. When the fruit is added the kefir ferments for a further 24 hours.

Makes 1 quart

1 quart Water Kefir (page 126)

Suggested flavorings
EITHER 3 slices fresh lime and 3 slices fresh lemon
OR 2 tablespoons blueberries and 2 tablespoons pomegranate seeds
OR 2 tablespoons fresh raspberries or 4 large strawberries, hulled and sliced

you will need a 1-quart glass Le-Parfait-style jar with a rubber seal, sterilized according to the instructions on page 11

Put all the ingredients in the glass jar and stir gently with a wooden spoon. Set aside to ferment on the kitchen countertop for 12 hours. Either drink it straight, as it is, or strain and drink. Best served chilled, though. You may like to sieve the fruit-flavored kefir before drinking, using a nylon sieve and pouring it into a clean pitcher. I like to strain it, pour it into a clean bottle, and keep it in the fridge. Drink within 24 hours.

Tip:

Avoid using metal utensils or containers when making kefir.

MILK KEFIR

MAKES 2 CUPS

2 tablespoons milk kefir grains (available online)
2 cups non-homogenized, organic whole milk

you will need a 2 pint glass Le-Parfait-style jar with a rubber seal, sterilized according to the instructions on page 11

Milk kefir is a fermented milk drink made from milk and milk kefir grains, which are different to water kefir grains. It is one of the most nutritious drinks you can find, as it's so rich in probiotics. Try drinking a small glass every day, or blending some milk kefir with fresh fruit to make smoothies. Use organic milk, raw if at all possible (see page 8).

Put the milk kefir grains and milk in the jar, stir well with a wooden spoon and close the jar. Place in the fridge for 12–24 hours while the kefir ferments. The kefir should thicken a little. Strain through a nylon sieve and drink as it is. As with the water kefir, start drinking a little at a time—no more than ½ cup.

You can keep reusing the milk kefir grains to make subsequent batches. They should multiply in quantity, which you can start to split and make more batches, or give away. Don't rinse them once they're fermented; use them as they are. Store in the fridge in a clean glass or plastic container.

COCONUT MILK KEFIR

MAKES 1¼ CUPS

2 tablespoons milk kefir grains (available online)
1 × 14 oz can of coconut milk

you will need a 2 pint glass Le-Parfait-style jar with a rubber seal, sterilized according to the instructions on page 11

This kefir is delicious made with coconut milk instead of cow's milk. Coconut milk can be fermented using milk kefir grains. It can be enjoyed as a thin, pouring yogurt for breakfast, or as a drink.

To use milk kefir grains to ferment coconut milk, you'll need to do a cow's milk ferment first with your milk kefir grains and repeat this process once every four ferments to keep your grains healthy and active. There's no need to rinse your milk kefir grains once they're fermented; use them as they are. The milk kefir grains will eat the lactose in the milk, meaning that anyone suffering from lactose intolerance should be fine with this. If in doubt, you may be best sticking to a powdered starter culture to ferment coconut milk.

Put the milk kefir grains and coconut milk in the jar, stir well with a wooden spoon and close the jar. Place in the fridge to ferment for 12–24 hours. The coconut milk kefir should thicken slightly, and the milk kefir grains may multiply slightly. Strain through a nylon sieve and drink as it is. Start with a glass of up to ½ cup initially. Best served chilled.

KOMBUCHA

Kombucha is a delicious fermented sweet tea. It is made using a scoby (Symbiotic Colony Of Bacteria and Yeast), which can either be bought or passed on from a friend. The scoby looks unusual, but it produces the most delicious drink that is lightly effervescent and tastes of apples.

You will need a small amount of kombucha to start a batch, so this is a great recipe to do with your friends and share among one another. Scobies can be peeled in half, or cut into quarters and pieces passed on to make another batch.

Makes approx. 2.5 litres

2 heaped tablespoons black loose-leaf tea (I use English Breakfast)
1 cup organic cane sugar
1 quart boiling water
1 quart filtered water (see page 8)
1 scoby or scoby piece (available online)
1 cup Kombucha (available online, or from a friend)

you will need a 3-quart glass Le-Parfait-style jar with a rubber seal, sterilized according to the instructions on page 11

Put the tea, sugar, and boiling water in a pitcher and stir well to dissolve the sugar. Set aside to infuse and cool to room temperature.

Strain the tea into a 3-quart glass jar and add the remaining ingredients. Stir well with a wooden spoon and then fasten the lid.

Set aside to ferment on the kitchen countertop for 5 days, after which time the kombucha will smell appley and lightly vinegary, and look clearer and more orange in color. I prefer to drink the kombucha at the younger stage, after 5 days, however you can leave it to ferment for up to 2 weeks if you wish. You will find that the flavor will become progressively more vinegary and effervescent the longer the kombucha ferments. I recommend starting by drinking ½ cup (no more) of kombucha. Reserve 1 cup of the kombucha to make a second batch.

FLAVORED KOMBUCHA

Once you've perfected making kombucha, you can start experimenting with different flavors.

Makes 1 quart

1 quart fermented Kombucha (see above)

For the flavorings

EITHER 3 large hibiscus flowers
OR a few sprigs of fresh mint or lemon balm
OR 2 teaspoons dried chamomile leaves

Pour off 1 quart of kombucha into a clean glass jar and stir in the flavoring of your choice. Set aside to ferment on the kitchen countertop for 12 hours. Drink as it is, or strain. Best served chilled.

Tip:

You can leave the lid on or off the bottle as it ferments. It can be more effervescent if the lid is fastened.

KOMBUCHA JUICE DRINKS AND SMOOTHIES

If you enjoy kombucha and want to use it more widely in your diet, try adding some to smoothies. You can either substitute the milk for kombucha or use it to thin your drink, adding extra flavor and nutrition. Most people prefer the following drinks made with kombucha that has been fermented for 5 days.

Serves 1 (Makes about 1¼ cups)

½ cup Kombucha (page 133), ⅓ cup apple juice, 3 fresh mint leaves

½ cup Kombucha, 1 medium banana, 6 strawberries

½ cup Kombucha, 3 tablespoons carrot juice, 2 teaspoons fresh ginger

½ cup Kombucha, 1 medium banana, a handful of blueberries, ½ teaspoon peeled and grated fresh ginger

½ cup Kombucha, 1 large banana, ¾ cup raspberries

½ cup Kombucha, ⅓ cup fresh pineapple, ¼ cup peeled cucumber

To make, simply blend the ingredients together in a blender and enjoy immediately.

SWEET LASSI

This creamy, refreshing drink is widely drunk across India, Pakistan, and Bangladesh. The sweet version, below, is probably the most popular, but a salted version is widely drunk, too. (If you want to try this, simply add a pinch of salt and a pinch of toasted ground cumin seeds to the mixture instead of the sugar.)
This is a lovely way to enjoy homemade yogurt. If you don't have any, simply use organic, live yogurt to enjoy its probiotic benefits in your drink.

Serves 2

¼ cup half and half
½ cup whole milk
1 cup natural yogurt (preferably Homemade Whole-milk Yogurt on page 44)
2 teaspoons superfine sugar

Put all the ingredients together in a blender and process until smooth. Taste, and check the level of sweetness is to your taste. You may wish to add a little more sugar. Serve immediately in chilled glasses, with a straw.

VARIATIONS

Once you have perfected the basic recipe, why not try out some different combinations:

Mango Lassi: Add 1 cup mango flesh.

Strawberry Lassi: Add 1⅓ cups fresh strawberries.

Almond Lassi: Add ¾ cup almond milk instead of the cream and milk.

Masala Lassi: Add 3 seeds from a green cardamom pod, a twist of black pepper, a strand of saffron, and a small grating of nutmeg.

MEAD

Mead is a very traditional honey wine. It is one of the easiest ferments to make and tastes wonderful. Unlike other homebrew drinks, you don't need special equipment or skill—a simple jar with a rubber seal will do.
In order for the sugar in the honey to ferment into alcohol, the mix needs to be free of chemicals, so do seek out raw honey, preferably local, to use in your brew.

MAKES 3 CUPS

3 cups filtered cold water (see page 8)
½ cup raw honey

you will need a 1-quart glass Le-Parfait-style jar with a rubber seal, sterilized according to the instructions on page 11

Pour the water into the jar, add the honey, and stir vigorously with a wooden spoon to dissolve the honey. Rest the lid on the jar, but don't clip it shut as you want a bit of air to circulate.

Set aside to ferment for at least 10 days on the kitchen countertop or in the pantry, stirring every day. The mead will look golden in color and smell sweet. You can drink the mead after 10 days, but you can also leave it for another 2–3 weeks to ferment, when it's a bit stronger. Mead is best enjoyed in small serving of perhaps ½ cup in a glass. The alcohol content can vary quite a bit—often between 7 and 18%, so do go easy, in case it's the latter.

The mead will keep unopened for up to 3 months. Once opened, store in a cool, dark place and consume within three weeks.

PRESERVES

HOMEMADE VINEGARS

Homemade vinegars are a lovely ingredient to have in the kitchen. Most of us don't think twice about the vinegars we use in cooking, but it's worth paying attention to what you use because good-quality vinegar gives a much better flavor.

To make red or white wine vinegar, you'll need to use some decent wine to ferment with a culture called an acetobacter, which converts the alcohol in the vinegar into acetic acid. Acetobacter is found in the "mother," which is the yeasty substance found at the bottom of bottled vinegars.

The process of making vinegar is fairly simple: you pour some white or red wine into a jar or crock, add some live vinegar, and then cover it over with a cheesecloth. And then you wait until it smells sweet and vinegary, about 2 months, at which point you can bottle it. All very simple in theory, however, in practice, it can be quite difficult to judge when the vinegar is ready; and you need good wine to make good vinegar. In my view, you're probably better off buying good-quality single varietal vinegar instead. Not only will this be easier, but also it will probably work out cheaper as well.

APPLE CIDER WINE VINEGAR

Makes 1.5 quarts

1 good-quality bottle of white or red wine
3 cups organic cider vinegar "with the mother"

you will need a 1.5-quart glass Le-Parfait-style jar with a rubber seal, sterilized according to the instructions on page 11

Tips:

Try flavoring your vinegar, with fresh herbs, for example, by adding a large sprig of tarragon to apple cider or white wine vinegar, which is very handy for making béarnaise sauce.

Keep tasting the vinegar as the flavor will keep changing—it's a living thing. Once you're happy with the flavor, bottle the vinegar and enjoy.

Vinegar is an incredibly useful ingredient to keep in your kitchen pantry. A tablespoon added to a slow-cooked dish can give the flavors a lift and add another dimension. It is also excellent used in mayonnaise and vinaigrette dressings for salads—just use three parts oil and one part vinegar, and whisk together.

Apple cider vinegar "with the mother" is widely available from health food stores, and combining this with wine makes for a lovely vinegar. In order for the wine to ferment, you need to add vinegar with the mother, which is "live." It can often look quite cloudy, with a sediment in the bottom of the bottle, and sometimes a thin jellylike layer floating on top.

Pour the wine and vinegar into the jar, cover the surface with cheesecloth and tie in place with string. Rest the lid on top, but don't fasten it because you need to allow the air to circulate. Set aside in a cool, dark place for 2 months before tasting. It should taste sweet (well, for vinegar) and well-rounded, but not excessively sharp and acidic. If you find the flavor too sharp, set it aside for an additional 2 weeks.

Store in a sealed container in a cool dark place for up to 6 months. Once opened, use within 3 months. The flavor will continue to develop over time.

BLACKBERRY VINEGAR

Blackberry vinegar is a gorgeous ingredient to use in dressings, or to add to slow-cooked meat dishes for an extra dimension of flavor. This sweetened vinegar is really versatile and makes a great gift. I love to use foraged blackberries for this, as wild fruit contains so much more flavor than cultivated.

Makes 2 cups

1¼ cups organic apple cider vinegar, or homemade vinegar (see page 142)
2 cups organic granulated cane sugar
3½ cups blackberries, rinsed

you will need a 1-quart glass Le-Parfait-style jar with a rubber seal, sterilized according to the instructions on page 11

Combine the vinegar and sugar in a medium saucepan and stir gently over low heat until all of the sugar has dissolved. Set aside to cool completely.

Place the blackberries inside the jar and crush them lightly with a fork. Pour the cold, sweetened vinegar over the fruit and close the lid.

Set aside in a cool, dark place for 2 weeks, after which time the vinegar will smell sweet and fruity. Strain the vinegar into a bottle and fasten the lid.

The vinegar will keep for up to 9 months.

CHERRIES IN VINEGAR

This is a divine way of preserving fresh cherries in white wine vinegar. They are fantastic eaten with cold meats, such as pâté and terrines, or with a cheeseboard.

Makes 1 × 1-quart jar

3 cups granulated cane sugar
$1\frac{2}{3}$ cups white wine vinegar, or homemade vinegar (see page 142)
1 cinnamon stick
2 cloves
1 star anise
4 cups whole fresh cherries

you will need a 1-quart glass Le-Parfait-style jar with a rubber seal, sterilized according to the instructions on page 11

Put the sugar, vinegar, and spices in a medium saucepan and stir over low heat until the sugar dissolves. Increase the heat and boil rapidly, until the liquid has reduced by one-third. Remove the pan from the heat, add the cherries, and set aside to cool.

Pour the contents of the pan into a glass jar, close the lid, and set aside in a cool, dark place for at least 2 weeks before opening.

Unopened, the cherries will keep for 6–12 months. Once opened, store in the fridge and use within a month.

CUCUMBER PICKLE

This Scandinavian pickle is fresh and fragrant. Perfumed with fresh dill, which is so prevalent in Nordic cuisine, it is superb eaten with salmon, smoked fish, rillettes and cheese—an extremely handy standby to have in the fridge.

MAKES 1 × 1-QUART JAR

1¼ cups granulated cane sugar
2 cups organic white wine vinegar or homemade vinegar (see page 142)
3 large cucumbers, thinly sliced (use a mandolin if you have one)
1 small white onion, very thinly sliced
1 teaspoon sea salt
5 large sprigs of fresh dill

you will need a 1-quart glass Le-Parfait-style jar with a rubber seal, sterilized according to the instructions on page 11

Put all the ingredients in the glass jar and stir well.

Store for up to 1 month in the fridge.

SPICED PICKLED PEACHES

This is a lovely way to make a glut of fresh peaches in the height of the season last throughout the year. They are fantastic enjoyed with a cold selection of salads, meats, and cheeses.

MAKES 1 × 2-QUART JAR

- 3½ cups granulated cane sugar
- 1⅔ cups apple cider vinegar, or homemade vinegar
- ½ teaspoon whole cloves
- 3 star anise
- 1 large cinnamon stick
- 1 teaspoon finely grated fresh ginger
- 2½ lbs peaches, pits removed and thickly sliced

you will need a 1-quart glass Le-Parfait-style jar with a rubber seal, sterilized according to the instructions on page 11

Put the sugar, vinegar, spices, and ginger in a medium saucepan and stir over low heat until the sugar has dissolved. Increase the heat and boil rapidly, until the liquid has reduced by one-third. Remove the pan from the heat, add the peaches, and set aside to cool.

Pour the contents of the pan, once cooled, into a glass jar. Ensure the peaches are completely submerged in the liquid, close the lid and set aside in a cool, dark place for 2 weeks before opening.

Unopened, the peaches will keep for 6–12 months. Once opened, store in the fridge and eat within a month.

FERMENTED JALAPEÑOS

A great ingredient to have on hand for anyone who enjoys jalapeños, these can be assembled in minutes and only take a week to ferment. An essential component in many Mexican dishes, these are fabulous served with scrambled eggs and chili con carne.

Makes 1 × 1-quart jar

2 cups jalapeño peppers, sliced into rings approx. ¼ in thick
approx. 3 cups filtered water (see page 8)
2 teaspoons sea salt
2 tablespoons Water Kefir (page 126)

you will need a 1-quart glass Le-Parfait-style jar with a rubber seal, sterilized according to the instructions on page 11

Put all the ingredients in the jar, pressing down well to ensure that the jalapeño slices are fully submerged. (If necessary, you may need to adjust the quantity of water.)

Seal the jar and set aside to ferment in a cool place, out of direct sunlight for 1 week.

These can be stored for up to 6 months, providing the jalapeños are fully submerged in the liquid. Once opened, store in the fridge and use within a month.

PRESERVED LEMONS

Preserved lemons are widely enjoyed across North Africa. They're an essential ingredient in many Moroccan and Egyptian dishes, adding a unique salty and fruity tang. It is very easy to make your own preserved lemons, and the delicious results make it very worthwhile. The key is to choose the best lemons you can find, and to use good-quality sea salt.

MAKES 1 × 1-QUART JAR

10 unwaxed lemons
6 tablespoons sea salt

you will need a 1-quart glass Le-Parfait-style jar with a rubber seal, sterilized according to the instructions on page 11

TIP:

You may wish to rinse the lemons before use, to remove some of the excess salt, but this is not essential.

Squeeze the juice from 4 of the lemons and pour into a pitcher, discarding the skins. Cover with plastic wrap and set aside in the fridge until needed.

Wash the 6 remaining lemons. Cut each lemon lengthwise into quarters, making sure you don't cut all the way through as you'll be stuffing the lemons with salt in the next stage.

Taking one lemon at a time, stuff the cavity of each lemon with 1 tablespoon of salt. Arrange the lemons inside the jar, pressing them down really well so they are packed tightly together. Close the lid.

Set aside to ferment on the kitchen countertop for an initial 3–4 days, by which time some of the juice will have run out of the lemons and started to fill the base of the container. Press down on the lemons to compact them further, and then pour in the reserved lemon juice to cover the lemons completely. (If they're not fully submerged, you will have to add some more lemon juice.)

Close the lid and set aside to ferment for an additional 4 weeks in a cool place, away from direct sunlight.

Unopened, the lemons will keep for up to a year in a cool dark place. Once opened, store in a cool place and consume within 6 months.

QUICK MANGO CHUTNEY

This super-fast recipe for mango chutney needs no maturing, so it can be eaten the same day. It keeps well, too, and is delicious with all manner of curries, meat and cheese plates, and even sandwiches.

Makes 1 × 1-quart jar

2 tablespoons olive oil
2 small white onions, thinly sliced
4 garlic cloves, peeled and finely chopped
1 in piece of fresh ginger, peeled and finely grated
2 large ripe mangoes, peeled, pit removed and flesh cut into 1 in cubes
6 tablespoons granulated cane sugar
6 tablespoons apple cider vinegar, or homemade vinegar (see page 142)
sea salt

you will need a 1-quart glass Le-Parfait-style jar with a rubber seal, sterilized according to the instructions on page 11

Heat the olive oil in a medium saucepan over medium heat. Add the onions, garlic, and ginger and cook for 5 minutes, until soft and fragrant. Add the mango, sugar, and vinegar and cook for an additional 10 minutes, until thick. Season the chutney with salt and set aside to cool before bottling.

This will keep for up to 3 months in a cool dark place. Once opened, store in the fridge and eat within 3 weeks.

INDEX

T

V

W

Y

Z

SUPPLIERS

CULTURES

This includes kefir grains, scobies, and sourdough levain. Once you get started with fermenting, you will have plenty of starter culture within a few weeks, as the cultures multiply over time. Therefore, you are likely to have enough to start giving some to friends. This is the best way to start, if you can. Otherwise, they are all available online, either through fermented cultures exchange groups online, or through online retailers, such as:

Happy Kombucha **http://happykombucha.co.uk**
Cultures For Health **http://www.culturesforhealth.com**
Sourdough school and online store **http://www.sourdo.com**
Jars **http://www.leparfait.com**

MISCELLANEOUS EQUIPMENT

For more jars, cheesecloth, sieves, bowls, jugs, bottles, string and digital thermometers:
http://www.surlatable.com
http://www.williams-sonoma.com

INGREDIENTS

I always recommend finding a good local supplier, such as a farm shop or health food store. The USDA offers a list of farmer's markets, which are great way to find local, often organic, produce in your area:
http://search.ams.usda.gov/farmersmarkets/

OTHER FERMENTED FOOD AND DRINK SUPPLIERS

Chocolate **http://www.rococochocolates.com**
Charcuterie **http://www.trealyfarm.com**
Olives **http://www.olivesetal.co.uk**
Coffee **http://www.monmouthcoffee.co.uk**
Beer **http://www.beerhawk.co.uk/**
Wine, including Organic, natural wines **http://www.bbr.com**

ACKNOWLEDGMENTS

This has been such a fun book to write and I have loved every minute of the process, from start to finish. It would not have happened without the guidance, encouragement, and backing from so many people, and I am grateful to you all.

First of all, I have to thank Darina, Tim, Penny, and Emer at Ballymaloe Cookery School for sparking my interest in fermentation, for sharing your knowledge, and for your support. This book wouldn't have happened without you.

Thanks are due to my brilliant agent, Clare, for your constant support and ability to make things happen.

I owe a huge debt of gratitude to Kyle for backing this book, and for doing so with such enthusiasm. It is a priviledge to work with you. I feel extremely fortunate to have worked with such a great team at Kyle Books, including Judith, Julia, Claire, Hannah, Marta, Mette, and Martina. Thank you for being such a pleasure to work with, and for really getting behind me.

It has been a complete joy to work with my Editor, Vicky. Thank you so much for getting what I wanted from the book straight away, making the whole experience run so smoothly, and being such a pleasure. Thank you for being so relaxed about my many changes, and my inability to work through edits with any speed whatsoever.

The creative team behind this book have been extraordinary. It was such a great pleasure to work with each one of you. Tara has been the best photographer I could have hoped for. Thank you for creating so many beautiful images, for being so lovely, and for your extraordinary levels of patience when it came to the author shots! To Annie, my unbelievably talented stylist on this book. I can't thank you enough for going the extra mile, for bearing with my corrections, and for cooking some seriously gorgeous food. Tabitha sourced the most beautiful props with such dedication and care; thank you. You have the most exquisite taste. Thanks too, to Sue and Lola, for all your help on shoots and to Georgina for taming my hair beautifully. It's been such fun working with you all.

I have a great team around me who help me to do what I do and I am extremely grateful for all that you do. Thank you to Louise for being the best kitchen companion I could ask for—always keeping me smiling AND doing the washing up!

Huge thanks to Lucy and Andy for doing so much for me, week in, week out, tolerating my crazy schedule, and the chaos in my wake. I appreciate it all.

Many thanks are due to Jean and Tony for your help, encouragement, props, ingredients, and reading material.

To my friends, thank you for being my biggest fans and for bearing with me whilst my work is all consuming. It means the world to me.

I can't thank my parents enough for your constant support. I know you continue to do an awful lot for me, more than I expect, and I am so grateful for it all.

Last but not least, thank you to Tony. For being you, and all that you do for me.

1858